PROCEEDINGS OF SYMPOSIA
IN APPLIED MATHEMATICS
Volume XXIII

MODERN STATISTICS: METHODS AND APPLICATIONS

AMERICAN MATHEMATICAL SOCIETY
PROVIDENCE, RHODE ISLAND
1980

LECTURE NOTES PREPARED FOR THE

AMERICAN MATHEMATICAL SOCIETY SHORT COURSE

MODERN STATISTICS:
METHODS AND APPLICATIONS

HELD IN SAN ANTONIO, TEXAS
JANUARY 7–8, 1980

EDITED BY
ROBERT V. HOGG

The AMS Short Course Series is sponsored by the Society's Committee on Employment and Educational Policy (CEEP). The Series is under the direction of the Short Course Advisory Subcommittee of CEEP.

Library of Congress Cataloging in Publication Data

American Mathematical Society Short Course on Modern Statistics: Methods and Applications, San Antonio, 1980.

Modern statistics, methods and applications.

(Proceedings of symposia in applied mathematics; v. 23)

"Lecture notes prepared for the American Mathematical Society Short Course [on] Modern Statistics: Methods and Applications, held in San Antonio, Texas, January 7–8, 1980."

Includes bibliographies.

1. Mathematical statistics–Congresses. I. Hogg, Robert V. II. American Mathematical Society. III. Title. IV. Series.

QA276.A1A45 1980 001.4'22 80-16093

ISBN 0-8218-0023-χ

1980 *Mathematics Subject Classification.* Primary 62D05, 68G10, 62G05, 62M10, 62F35.

CONTENTS

CONTENTS

[illegible]

[illegible]

[illegible]

[illegible]

[illegible]

[illegible]

PREFACE

This volume contains the lecture notes prepared by the speakers for the American Mathematical Society Short Course on *Modern Statistics*: *Methods and Applications* given in San Antonio, Texas, on January 7–8, 1980.

We were very pleased with the substantial attendance at the course. The skills of the lecturers and the enthusiasm of the participants encouraged the AMS Committee on Short Courses to request that these notes be published. We are indebted to our colleagues for this support and the AMS office for the cooperation in publishing these proceedings.

Of course, the choice of topics from a field as large as Statistics is a difficult one. However, I did want to avoid any substantial overlap with the highly successful short course on statistics held three years earlier in St. Louis, January, 1977. Therefore it seemed very natural to begin with one important topic that is sometimes overlooked in an introductory course, particularly one in mathematical statistics. Yet this topic is one through which the general public most often hears about statistics, namely, survey sampling. Wayne Fuller spoke on "Samples and Surveys", noting the operations necessary in conducting a survey of a human population. In his article, he explains the construction of a probability sample design and the corresponding optimal estimators.

The more general problem of the design and analysis of an experiment was covered by Peter John in his "Analysis of Variance". These techniques have, for years, been extremely important in applications and have also motivated a large amount of statistical research. It is clear that even in an elementary design the experimenter must understand the importance of randomization.

Nonparametric statistical methods have played a major role in modern statistics. Two coordinated talks on that subject were given by Ronald Randles and Thomas Hettmansperger. Randles introduced distribution-free rank tests, such as one by Wilcoxon, and some of their good asymptotic properties. Hettmansperger then explained how these rank tests could be used to obtain point and interval estimates for various parameters, including the regression situation. These resulting

R-estimates are very robust because they are not highly sensitive to reasonable deviations from the underlying assumptions.

The important topic of regression was continued by considering isotonic regression and time series. F. T. Wright showed how to use the method of maximum likelihood to estimate ordered parameters. Then Douglas Martin considered a time sequence of data. After presenting a collection of interesting examples, he discussed appropriate models and their estimates, including robust ones.

While it is impossible to cover all of Statistics in six articles, these and their references should prove useful to those who wish to learn something of the natures of modern statistics. In that regard, I must also call your attention to *Studies in Statistics* that I had the opportunity to edit for Volume 19 of Studies in Mathematics under the sponsorship of the Mathematical Association of America. I hope that this present volume, along with that one, will provide the interested reader a good introduction to modern statistical methods.

Robert V. Hogg
University of Iowa
March, 1980

Proceedings of Symposia in Applied Mathematics
Volume 23
1980

SAMPLES AND SURVEYS

Wayne A. Fuller[1], Iowa State University

I. Introduction.

The design and execution of a large scale survey is a sizeable research undertaking. We outline the steps in such an operation.

A. Definition of the objectives.

B. Specification of the procedures.

   1. Universe of interest.
   2. Data to be collected and method of collection.
   3. Sample design.
   4. Questionnaire design.

C. Field work.

D. Data processing.

   1. Coding.
   2. Editing.
   3. Estimation and tabulation.

E. Report preparation.

We assume that the objectives of the study require obtaining data from an existing group of elements. The <u>universe</u> is the collection of elements about which statements will be made. In most surveys data are collected on a large number of characteristics. The regular polls (Gallup, etc.) record items such as age, race, sex, place of residence, and political affiliation, in addition to responses on a few questions of current interest.

The reasons for observing a part of the universe (taking a sample) instead of the entire universe (a census) are all practical. First the research budget seldom permits observing every element of the population. A personal interview now costs on the order of $30 to $100 to complete. Also, in certain quality testing situations, the observations are destructive. It is of little use to know that a lot of light bulbs will last an average of

---

1980 Mathematics Subject Classification 62D05.
[1]This research was partly supported by Joint Statistical Agreement JSA 79-10 with the U.S. Bureau of the Census.

200 hours if they have all been burned to establish that fact.

There are other disadvantages of censuses. The first is timing. The data for the 1970 Census of Population were collected beginning in April 1970. The first preliminary raw count reports (for states, counties and municipalities) were available in May through October 1970. The advanced reports become available in the period September 1970 through February 1971. The U.S. summary report was released in January 1972.

More subtle is the problem of quality control in a census. The population census of the United States requires over one quarter million field workers and supervisors. Because fewer interviewers are required for a sample, it should be possible to select better people and to better supervise the field operation.

Once one has decided that a census is impossible, the questions become: What kind of sample? How large a sample?

A sample is a portion of the universe. A random sample (or probability sample) is a sample selected in such a way that the probability of selecting every sample is known.

A simple random sample is a sample of n elements chosen from a population of N elements in such a way that each one of the ${}_N C_n$ samples has an equal probability of being selected.

A purposive sample (an alternative term is judgmental sample) is any sample that is not a probability sample. Generally speaking, purposive samples are selected to meet certain criteria. The prime example is the political subdivision that has voted for the winner in the last ten elections.

What is the place of the two kinds of sampling? Let us first consider the problem from an empirical point of view. An experiment cited by Jessen (1978, p. 18) compared two methods of sampling a universe of 126 stones. Members of a statistics class were instructed to look at the entire universe and then to select a sample that would represent the average weight of the stones. The sixteen students selected three samples of sizes 1, 2, 5 and 10 and one sample of size 20. Simple random samples of the same sizes were also selected. (126 samples of size one, 30 of size 2, 90 of size 5, 60 of size 10 and 10 of size 20 were selected.)

Table 1. Mean absolute deviation for two types of sample selection.[1]

Type of Sample	Sample Size				
	1	2	5	10	20
Judgment	40.0	44.9	35.3	38.5	31.0
Random	80.6	71.4	41.3	34.1	26.2

[1]From Jessen (1978, p. 18)

The conclusion of the Jessen experiment seems a part of scientific practice. That is, you can expect that your journal article will be accepted if you are working with a very small judgment sample, on the order of 5 or less, but can anticipate difficulties if you submit an article based on a large judgment sample.

II. Simple Random Sampling.

We present a few of the results on simple random sampling. Because of the simplicity of these results, sampling offers an excellent method of introducing a student to statistics. The population of possible samples can be enumerated and the expectation of a random variable can be introduced as the average over the finite number of possible outcomes.

Let the population be composed of N elements. Let the value of the y-characteristic of the elements be denoted by $\{y_i: i=1, 2, \ldots, N\}$. The probability that a particular element appears in a simple random sample is n/N. The probability that a particular pair of elements appears in the sample is $[N(N-1)]^{-1} n(n-1)$. From these basic properties of simple random samples several results are immediate.

RESULT 1. The sample mean is unbiased for the population mean.

RESULT 2. The variance of the sample mean is

$$E\{(\bar{y}-\bar{Y})^2\} = \frac{N-n}{N-1}\frac{\sigma^2}{n} = \frac{N-n}{N}\frac{S^2}{n}, \tag{1}$$

where $S^2 = N(N-1)^{-1}\sigma^2$, $\sigma^2 = N^{-1}\Sigma_{i=1}^{N}(y_i-\bar{Y})^2$ and $\bar{y}$ is the sample mean.

RESULT 3. An unbiased estimator of S^2 is

$$s^2 = (n-1)^{-1}\sum_{i=1}^{n}[y_i-\bar{y}]^2 . \tag{2}$$

Assume that the characteristic y takes on the two values 0 and 1. Let N_1 of the elements be ones and $N-N_1$ of the elements be zeros. If a simple random sample of size n is selected from the N elements, the probability that exactly n_1 of the elements will possess a y-charcteristic of 1 is

$$P\left(\sum_{i=1}^{n} y_i = n_1\right) = \frac{\binom{N_1}{n_1}\binom{N-N_1}{n-n_1}}{\binom{N}{n}} \tag{4}$$

Table 2 contains the probabilities for all possible values of (N_1, n_1) for a sample of $n=5$ selected from a population of $N=15$. Two lines have been drawn through the values. The lines are such that the sum of the probabilities to the right of the right line in every row is less than 0.12 .

The lines enable us to define an interval for each sample outcome such that, for each value of N_1, the probability is greater than 0.88 that the interval will cover the true N_1. The intervals for each n_1 are defined by the horizontal lines in Table 2. The intervals of Table 2 are the classical confidence intervals introduced by Neyman (1934, 1935). In particular, see Neyman (1934, p. 624).

Table 2. Probabilities of sample outcomes for samples of size five selected from fifteen

Number of 1's in Population N_1	Number of 1's in Sample n_1: 0	1	2	3	4	5
0	1.0000	0	0	0	0	0
1	0.6667	0.3333	0	0	0	0
2	0.4286	0.4762	0.0952	0	0	0
3	0.2637	0.4945	0.2198	0.0220	0	0
4	0.1538	0.4395	0.3297	0.0733	0.0037	0
5	0.0839	0.3497	0.3996	0.1499	0.0166	0.0003
6	0.0420	0.2517	0.4196	0.2398	0.0449	0.0020
7	0.0187	0.1632	0.3916	0.3263	0.0932	0.0070
8	0.0070	0.0932	0.3263	0.3916	0.1632	0.0187
9	0.0020	0.0449	0.2398	0.4196	0.2517	0.0420
10	0.0003	0.0166	0.1499	0.3966	0.3497	0.0839
11	0	0.0037	0.0733	0.3297	0.4395	0.1538
12	0	0	0.0220	0.2198	0.4945	0.2637
13	0	0	0	0.0952	0.4762	0.4286
14	0	0	0	0	0.3333	0.6667
15	0	0	0	0	0	1.0000

Table 3. Possible outcomes for a population with $N_1 = 8$

n_1	Probability	Interval	Statement
0	0.0070	[0, 4]	Wrong
1	0.0932	[1, 7]	Wrong
2	0.3263	[3, 10]	Right
3	0.3916	[5, 12]	Right
4	0.1632	[8, 14]	Right
5	0.0187	[11, 15]	Wrong

The possible outcomes for $N_1=8$ are given in Table 3. The probability of a wrong statement is 0.1189 and the probability of a correct statement is 0.8811.

The confidence interval is particularly forceful when used with random sampling of zero-one characteristics. This is because the model is guaranteed to be correct. The use of randomization in sample selection creates the population of samples for which our statement applies. In random sampling from a finite universe, we know the nature of the sampling distribution because we have created it!

To establish confidence intervals for the mean of a finite population for a characteristic that is not 0-1 we must expand our theoretical constructs. One approach is to assume that the finite population is a random sample from an infinite superpopulation. For example, assume that the finite population is a random sample from a normal population. If we select a random sample from the finite population, we have created two independent random samples, one of size n and one of size $N-n$, from the original population. Then the difference $\bar{y}-\bar{Y}$, where $\bar{y}$ is the mean of the n elements and $\bar{Y}$ is the mean of the N elements, is distributed as a normal random variable with mean zero and variance

$$V\{\bar{y}-\bar{Y}\} = \frac{N-n}{Nn}\sigma^2 .$$

It follows that

$$t = [(Nn)^{-1}(N-n)s^2]^{-\frac{1}{2}}(\bar{y}-\bar{Y})$$

is distributed as Student's t. The denominator of the t statistic is an unbiased estimator of the variance conditional on $(y_1, y_2, \ldots, y_N)$ as well as an unbiased estimator of the unconditional variance. In this argument the distribution of $\bar{y}-\bar{Y}$ is for the population of all possible pairs of samples of size n and $N-n$ selected from the parent population. Perhaps, because samplers have traditionally preferred to think of the finite population as fixed, this theoretical construct seldom appears in sampling texts.

In defining a sequence for a central limit theorem for sampling from a finite universe, one must consider a sequence of samples selected from a sequence of populations. To obtain a limiting normal distribution, the sequence of populations must satisfy certain conditions. Two approaches have been used. One is to specify conditions on the population sequence itself. In this case the distribution is for the population of random samples created by randomization conditional on fixed population values. Madow (1948) and Hájek (1960) give results of this type. The second approach is to assume that the finite population is a sample from an infinite population. We state a result of the second type.

THEOREM. Let $\{\mathcal{U}_t : t=1, 2, \ldots\}$ denote a sequence of finite populations, where $\mathcal{U}_t$ is a random sample of size N_t, $N_t > N_{t-1}$, selected from an infinite population. Assume the infinite population possesses finite first and second moments. Let simple random samples of size n_t, $n_t > n_{t-1}$ be selected from N_t. Let

$$\lim_{t \to \infty} N_t^{-1} n_t = f, \qquad 0 \le f < 1 .$$

Then

$$n_t^{\frac{1}{2}} (\bar{y}_t - \bar{Y}_t) \xrightarrow{\mathcal{L}} N(0, (1-f)\sigma^2) ,$$

where $\bar{y}_t$ is the sample mean and $\bar{Y}_t$ is the population mean for population t and σ^2 is the variance of the infinite population.

III. Unequal Probability Sampling.

To use probability ideas in sampling, it is not necessary that each element have an equal probability of entering the sample. To introduce the ideas of unequal probability sampling, consider a population of five elements with characteristics $\{y_1, y_2, y_3, y_4, y_5\}$. Assume that we create ten slips of paper as given in Table 4. Let p_i denote the fraction of the slips that have element i recorded on them. Then the average over the ten slips of the ratios $p_i^{-1} y_i$ is $\Sigma_{i=1}^5 y_i$. Therefore, if we randomly choose one slip, an unbiased estimator of the total of y is $p_i^{-1} y_i$. If we use replacement sampling, the average of the n values we observe is an unbiased estimator of the population total Y with variance

$$V\{\hat{Y}\} = n^{-1} \sum_{i=1}^{N} p_i (p_i^{-1} y_i - Y)^2 . \tag{8}$$

In nonreplacement sampling, the probabilities of selection π_i are typically specified so that $\Sigma_{i=1}^N \pi_i = n$, where n is the number of elements to be included in the sample. With this normalization, the unbiased estimator of the population total is

$$\hat{Y} = \sum_{j=1}^{n} y_j \pi_j^{-1} .$$

The variance of the estimator depends upon the joint probabilities of selection π_{ij};

$$V\{\hat{Y}\} = \sum_{i=1}^{N} \pi_i^{-1} (1-\pi_i) y_i^2 + \sum_{i \neq j}^{N} \pi_i^{-1} \pi_j^{-1} (\pi_{ij} - \pi_i \pi_j) y_i y_j . \tag{9}$$

An unbiased estimator of this quantity is

$$\hat{V}\{\hat{Y}\} = \sum_{i<j}^{n} \pi_{ij}^{-1} (\pi_i \pi_j - \pi_{ij})[\pi_i^{-1} y_i - \pi_j^{-1} y_j]^2 . \qquad (10)$$

Table 4. Population of slips for unequal probability sampling.

Original Population Element	y-value	Number of Such Slips	$p_i^{-1} y_i$
1	y_1	4	2.5 y_1
1	y_1	4	2.5 y_1
1	y_1	4	2.5 y_1
1	y_1	4	2.5 y_1
2	y_2	3	(10/3) y_2
2	y_2	3	(10/3) y_2
2	y_2	3	(10/3) y_2
3	y_3	1	10 y_3
4	y_4	1	10 y_4
5	y_5	1	10 y_5

IV. Sample Sizes and Sample Frames.

The first question a consulting survey statistician hears from the client is: How many ... do I need? The question, formulated so that an answer is possible, requires considerable information:

(1) A statement of desired closeness for the final answer. For example: "I wish my estimate of the mean of y to be within d units of the true value with probability $1-\alpha$."

(2) An estimate of the variability in the parent population. For example: "The variable y is similar to the variable x which has a variance of σ^2."

Assume the existence of an idealized client that specifies that the estimate of the proportion is to be within 0.02 of the true proportion with probability $1-\alpha$. Some required sample sizes are given in Table 5.

To select a probability sample, it is necessary to create a list, called the sampling frame, such that every element of the universe is associated with at least one item on the list. We consider frames such that each element is associated with one and only one item on the list. The frame may be a physical list such as a list of automobile registrations or it may be a conceptual list such as the list of all possible latitude, longitude coordinates for every point of the land area in the United States.

Table 5. Size of sample required for observed proportion p to be within 0.02 of population proportion P with at least probability $1-\alpha$.

Population Size	True Proportion $P = 0.5$		$P = 0.1$
	$1-\alpha = 0.95$	$1-\alpha = 0.99$	$1-\alpha = 0.99$
100	98	99	98
1,000	720	816	610
2,500	1,226	1,559	948
10,000	1,922	2,946	1,304
25,000	2,206	3,571	1,424
50,000	2,306	3,844	1,465
100,000	2,359	3,996	1,482
∞	2,401	4,147	1,493

The construction of a standard sampling frame is the task of constructing a list of <u>primary sampling units</u> such that every element in the population is in exactly one of the primary sampling units. Some primary sampling units may contain no elements and some may contain several. Because so few lists of human and economic populations exist, it is often necessary to create a list of primary sampling units that can be identified in the field operation. One of the most important frames of this type is the <u>area frame</u>.

The area sample is an example of a <u>cluster sample</u>. If any primary sampling unit in the frame contains more than one or less than one observation unit (element) the primary sampling units are called <u>clusters</u>.

Cluster sampling is used for two reasons.

(a) It may be impossible or prohibitively expensive to construct a list of observation units.

(b) For a fixed expenditure, it is often possible to obtain a smaller mean square error for an estimator by observing groups of observation units.

The estimation formulas presented for simple random samples apply for cluster samples with y_i being the total of the y-characteristics of the elements in the primary sampling unit.

VII. Ratio and Regression Estimation.

Methods of estimation employed in survey sampling, beyond the basic mean-type estimators we have presented, are ratio and regression estimation. Assume that we have available some information about the population. As an example, consider a study of farms. We have very good data on the land in farms. Call the total land in farms X and assume it to be known. Assume, less realistically, that we draw a random sample of farms. Let the y-characteristic be the acres of corn. Then the ratio estimator of the total acres of corn is

$$\hat{Y}_r = \left(\sum_{i=1}^{n} y_i\right)\left(\sum_{i=1}^{n} x_i\right)^{-1} X = \bar{x}^{-1}\,\bar{y}\,X \,, \tag{21}$$

where x_i is the acreage of the ith sample farm. The simple regression estimator of the total acres of corn is

$$\hat{Y}_\ell = N[\bar{y} + b(\bar{X} - \bar{x})] \,, \tag{22}$$

where $\bar{X} = N^{-1} X$ and b is the usual least squares regression coefficient.

Neither of these estimators is unbiased. Defining a sequence of populations, it is possible to demonstrate that $n^{\frac{1}{2}}(\bar{y}_r - \bar{Y})$ is approximately distributed with mean zero and variance

$$(1-f)(S_Y^2 - 2RS_{XY} + R^2 S_X^2) \,, \tag{23}$$

where $f = N^{-1} n$, $R = \bar{Y}/\bar{X}$, and $\bar{y}_r = N^{-1}\hat{Y}_r$. Similarly, $n^{\frac{1}{2}}(\bar{y}_\ell - \bar{Y})$ is approximately distributed with mean zero and variance

$$(1-f)(S_Y^2 - B\,S_{XY}) \,, \tag{24}$$

where $B = S_X^{-2} S_{XY}$ and $\bar{y}_\ell = N^{-1}\hat{Y}_\ell$.

VIII. Survey Design.

The objective of survey design is to use the available information to create a method of sampling and an estimation rule that yields estimators with desirable properties. Some desirable properties of design-estimator pairs are:

1. Unbiasedness.
2. Accuracy. A measure of the accuracy of an estimator $\hat{\theta}$ of θ is the mean square error.

   $$\text{M.S.E.} = E\{(\hat{\theta} - \theta)^2\}$$

3. Consistency.
4. Scale invariance. The estimator $\hat{\theta}(\underset{\sim}{y})$ is scale invariant if $\hat{\theta}(k\underset{\sim}{y}) = k\hat{\theta}(\underset{\sim}{y})$ for all fixed k.
5. Location invariance. The estimator $\hat{\theta}(\underset{\sim}{y})$ is location invariant if

$\theta(k+\underset{\sim}{y}) = k + \theta(\underset{\sim}{y})$ for all fixed k .

6. Simplicity.
7. Internal consistency. The estimators $\hat{\theta}_1(\underset{\sim}{y})$, $\hat{\theta}_2(\underset{\sim}{y})$ and $\hat{\theta}_3(\underset{\sim}{y})$ are internally additively consistent for θ_1, θ_2, and θ_3 if

$$\hat{\theta}_1(\underset{\sim}{y}) + \hat{\theta}_2(\underset{\sim}{y}) = \hat{\theta}_3(\underset{\sim}{y})$$

for $\theta_1 + \theta_2 = \theta_3$.

8. Accurate internal estimator of the variability of the estimator.
9. Robustness. A sample design-estimator pair is robust if departures of the sampled population from that anticipated at the design stage produce small decreases in accuracy.
10. Practicality.

Godambe (1955) pointed out that if we pay attention to the individual values of the population (treat them as fixed constants) there is no probability design that is best for all possible populations. For a particular set of positive y's we can obtain zero variance by making the selection probabilities p_i of Table 5 proportional to y_i .

Godambe's result suggests that we should quantify the prior information associated with the individual elements of the population at the design stage. We use Godambe's formalism to present the design problem.

Let $\mathcal{U} = \{u_i: i=1, 2, \ldots, N\}$ denote the N units of the finite population. Let s denote a subset of the units of the population and let $\mathcal{S}$ be the set of all subsets, s. Let d be an estimator constructed from the units of s . Let p denote a sampling design. The sampling design assigns probabilities, summing to one, to the elements s of $\mathcal{S}$. Let $\mathcal{A}$ denote the prior information about $\mathcal{U}$ available at the time a survey design is to be chosen. Let $\mathcal{P}$ denote the set of all possible designs and $\mathcal{D}$ the set of all possible estimators. Given $\mathcal{A}$ and an optimality criterion, the problem of choosing the optimal $p \subset \mathcal{P}$ and $d \subset \mathcal{D}$ is the problem of survey design.

The class $\mathcal{D}$ is often restricted to the class of linear estimators. A linear estimator can be written as

$$d = \sum_{i=1}^{n} w_i(\underline{s})\, y_i \tag{25}$$

where the weights are permitted to be a function of the element identification and of the sample.

Let us consider a simple statistical problem. Let the observations $\{y_i: i=1, 2, \ldots, n\}$ be independently distributed (μ, σ^2) where μ and σ^2 are unknown. That is, $\{y_1, y_2, \ldots, y_n\}$ is a random sample selected from a parent distribution with mean μ and variance σ^2 . Then $\bar{y}$ is the best linear unbiased estimator of μ . Assume that a second sample of size $N-n$ is to be selected from the same distribution. Denote the mean of the second sample by $\bar{y}_{N-n}$. Then the best linear unbiased predictor of $\bar{y}_{N-n}$ is also $\bar{y}$. It follows that $\bar{y}_n$ is the best linear unbiased predictor of

$$\bar{Y} = N^{-1} [n\bar{y} + (N-n)\bar{y}_{N-n}] .$$

If we rearrange the timing of the sampling operations we can use this model as a framework for developing sample designs and estimators. Assume that we are first given the sample of N observations from the population. Then from the N we select a simple random sample of size n. The sample mean remains the best linear unbiased predictor of the mean of the remaining $N-n$ elements. Considering the finite population to be a sample from a superpopulation is a way to formalize our prior information about the finite population. We can conceptualize the problem in this way at the design stage even if we plan to treat the finite population as a set of fixed numbers at the ultimate estimation stage.

At the design stage we assume that the finite population is a vector selected from an infinite population with mean $\underset{\sim}{\mu}$ and nonsingular $N \times N$ covariance matrix $\underset{\sim}{\Sigma}$, where $\underset{\sim}{\Sigma}$ has typical element σ_{ij} . In later formulations $\underset{\sim}{\mu}$ and $\underset{\sim}{\Sigma}$ may be specified to be functions of unknown parameters.

We restrict designs to the class of random designs. In this formulation there are two sources of "variation" to be considered at the design stage. The first is associated with the superpopulation from which the finite population is viewed as a sample. The second is that introduced by the random sample design. We denote the expectation with respect to the superpopulation with the symbol $\mathcal{E}$ and the expectation with respect to the sampling design by E .

An estimator d is <u>conditionally model unbiased</u> for $\bar{Y}$ if

$$\mathcal{E}\{(d - \bar{Y}) | s\} = 0 . \tag{26}$$

The conditioning in (26) is with respect to the elements of s and the prior information for those elements, but not on the y-characteristics of s . The estimator d is said to be a <u>model unbiased predictor</u> of $\bar{Y}$ if (26) holds for all s in $\mathscr{S}$ such that $p\{s\} > 0$.

An estimator d is <u>unbiased for $\bar{Y}$ with respect to the design</u> p if

$$E\{d\} = \bar{Y} , \tag{27}$$

for all $(y_1, y_2, \ldots, y_N)$ contained in N-dimensional Euclidean space, where

$$E\{d\} = \Sigma_s d(s)\ p(s)$$

and Σ_s denotes the summation over all samples s. An estimator d satisfying (27) may also be called design unbiased or p-unbiased. The anticipated mean square error of an estimator d of $\bar{Y}$ is defined by

$$\text{AMSE} = \mathcal{E}\{E[(d - \bar{Y})^2]\} .$$

Model predictors, the concepts of design unbiasedness and the operators $\mathcal{E}$ and E are discussed by Cassel, Särndal and Wretman (1977, Ch. 4)

The anticipated variance of an estimator d of $\bar{Y}$ is the variance of the population of estimates created by the sample design and sampling from the superpopulation;

$$AV\{d - \bar{Y}\} = \mathcal{E}\{E(d - \bar{Y})^2]\} - [\mathcal{E}\{E(d - \bar{Y})\}]^2 .$$

Given a superpopulation model with finite moments and a design whose probabilities are independent of the y-values; we may interchange the order of the expectation operators to obtain,

$$AV\{d - \bar{Y}\} = E\{\mathcal{V}(d - \bar{Y}|s)\} + V[\mathcal{E}(d - \bar{Y}|s)] , \tag{28}$$

where

$$\mathcal{V}(d - \bar{Y}|s) = \mathcal{E}\{[(d - \bar{Y}) - \mathcal{E}(d - \bar{Y}|s)]^2|s\} , \tag{29}$$

$$V\{d\} = E\{[d - E(d)]^2\}$$

If d is unbiased for $\bar{Y}$ under the design the anticipated variance is the design expectation of the conditional model variance:

$$AV\{d - \bar{Y}\} = \mathcal{E}[V\{d\}] = \mathcal{E}[E\{[d - E(d)]^2\} . \tag{30}$$

If d is unbiased for $\bar{Y}$ under the model the anticipated variance is the design expectation of the conditional model variance:

$$AV\{d - \bar{Y}\} = E[\mathcal{V}\{d - \bar{Y}|s\}] . \tag{31}$$

We first consider the design-estimation problem for a population that has been divided into k groups called strata. Let y_{hj} denote the jth observation in the hth stratum.

THEOREM. Let the finite population be a realization from the superpopulation defined by

$$y_{hj} = \mu_h + e_{hj} , \qquad h = 1, 2, \ldots, k; \quad j = 1, 2, \ldots, N_h ,$$

$$E\{e_{hj}^2\} = \sigma_h^2 , \quad E\{e_{hj}\} = 0 ,$$

where the e_{hj} are independent random variables, σ_h^2 is known and $N = \Sigma_{h=1}^{k} N_h$. Restrict consideration to the class of linear design unbiased estimators and the class of fixed sample size designs that admit design unbiased estimators. Then the design-estimator pair with minimum anticipated variance for the finite population mean

$$\bar{Y} = N^{-1} \sum_{h=1}^{k} N_h \bar{Y}_h = N^{-1} \sum_{h=1}^{k} \sum_{j=1}^{N_h} y_{hj}$$

is the estimator

$$\bar{y}_{st} = N^{-1} \sum_{h=1}^{k} N_h \bar{y}_h , \tag{32}$$

where

$$\bar{y}_h = n_h^{-1} \sum_{j=1}^{n_h} y_{hj} ,$$

and the design that chooses a simple random sample of size n_{h*} from the hth stratum, with

$$n_{h*} = (\Sigma_{j=1}^{k} N_j \sigma_j)^{-1} N_h \sigma_h n . \tag{33}$$

PROOF. Consider a sample that has n_h sample elements in the hth stratum, all $n_h > 0$. Then, under the model, the conditionally best linear unbiased estimator of

$$\sum_{h=1}^{k} N^{-1} N_h \mu_h$$

is the estimator (32). This is also the conditionally minimum variance linear unbiased predictor of $\bar{Y}$. The model variance is

$$\mathcal{V}\{(\bar{y}_{st} - \bar{Y}) | s\} = N^{-2} \sum_{h=1}^{k} N_h n_h^{-1} (N_h - n_h) \sigma_h^2 . \tag{34}$$

The n_h that minimize (34) are the n_{h*} given in (33). (We ignore the modifications required to produce n_{h*} that are all positive integers.) To prove that there exists no superior linear design unbiased procedure we consider an alternative estimator

$$d_* = \bar{y}_{st} + b$$

where b is a linear estimator. For d_* to be a design unbiased estimator, we must have $E\{b\} = 0$. Then the anticipated variance of d_* is

$$\mathcal{E}[V\{d_*\}] = \mathcal{E}[V\{\bar{y}_{st} - \bar{Y}\} + 2 \operatorname{Cov}\{\bar{y}_{st} - \bar{Y}, b\} + V\{b\}].$$

Consider

$$\begin{aligned} \mathcal{E}[\operatorname{Cov}\{\bar{y}_{st} - \bar{Y}, b\}] &= \mathcal{E}[E\{(\bar{y}_{st} - \bar{Y})b\}] \\ &= E\{\mathcal{E}[(\bar{y}_{st} - \bar{Y})b | s]\} . \end{aligned}$$

Because $\bar{y}_{st}$ is conditionally the best linear model unbiased estimator, $\mathcal{E}\{(\bar{y}_{st} - \bar{Y})b|s\} = 0$ for any b that is linear. It follows that there is no linear design unbiased estimator and design with smaller anticipated variance than $\bar{y}_{st}$ with $n_h = n_{h*}$. ▯

We now consider a superpopulation model that leads to the regression estimator. Let the finite population be a realization from the model

$$y_i = \underset{\sim}{z}_i \underset{\sim}{\beta} + \epsilon_i \ , \tag{35}$$

$$\begin{aligned} E\{\epsilon_i \epsilon_j\} &= \gamma_{ii}\sigma^2 \qquad & i = j \\ &= 0 & i \neq j \end{aligned} \tag{36}$$

where $\underset{\sim}{z}_i = (\gamma_{ii}^{\frac{1}{2}}, 1, \gamma_{ii}, x_i)$ and $\underset{\sim}{\beta}' = (\beta_1, \beta_2, \beta_3, \beta_4)$. Assume $\underset{\sim}{\beta}$, and σ^2 are unknown and $\bar{X}_t$ and γ_{ii} are known.

In the study of estimators for such a population, we will consider a sequence of populations and samples. Let $\{y_j, \pi_j\}$ denote a vector sequence where $1 > \lambda_1 > \pi_j > \lambda_2 > 0$ for all j. Let $\{\mathcal{U}_t\}$ denote a sequence of finite populations of size $\{N_t\}$, $0 < N_1 < N_2 < N_3 \ldots$, created from the sequence $\{y_j, \pi_j\}$, where $\mathcal{U}_1$ is composed of the first N_1 elements of $\{y_j, \pi_j\}$, $\mathcal{U}_2 \supset \mathcal{U}_1$ is composed of the first N_2 elements of $\{y_j, \pi_j\}$, etc. Let a sequence of samples of size $n_t = \Sigma_{i=1}^{N_t} \pi_i$ be created from the sequence of populations. Let element i be included in the sample with probability π_i and let the joint probabilities for the tth population be denoted by $\pi_{ij(t)}$.

RESULT 4. (Godambe and Joshi (1965)) Assume model (35) with ϵ_i independent of ϵ_j. Then the minimum anticipated variance for design unbiased estimators is

$$M_t \ \sigma^2 = N_t^{-2} \ \sigma^2 \left[n_t^{-1} \left(\sum_{i=1}^{N_t} \gamma_{ii}^{\frac{1}{2}} \right)^2 - \sum_{i=1}^{N_t} \gamma_{ii} \right] . \tag{37}$$

RESULT 5. Let replacement samples of size n_t be selected. Let the probability of selecting element i on each draw be p_i and let p_i be proportional to $\gamma_{ii}^{\frac{1}{2}}$. Consider the class of linear design unbiased estimators of $\bar{Y}_t$ of the form

$$\begin{aligned} n_t^{-1} \sum_{i=1}^{n_t} (N_t \ p_i)^{-1} \, [y_i - C_2(1 - N_t p_i) - C_3(\gamma_{ii} - N_t p_i \ \bar{G}_t) \\ - C_4(x_i - N_t p_i \ \bar{X})] \end{aligned} \tag{38}$$

with (C_2, C_3, C_4) fixed. The design variance of the estimator is minimized by setting (C_2, C_3, C_4) equal to the last three elements of

$$\underset{\sim}{C}_t = (\underset{\sim}{Z}'_N \underset{\sim}{P}_N^{-1} \underset{\sim}{Z}_N)^{-1} \underset{\sim}{Z}'_N \underset{\sim}{P}_N^{-1} \underset{\sim}{y}_N , \tag{39}$$

where

$$\underset{\sim}{Z}'_N = (\underset{\sim}{z}'_1, \underset{\sim}{z}'_2, \ldots, \underset{\sim}{z}'_N) , \qquad \underset{\sim}{P}_N = \text{diag}(p_1, p_2, \ldots, p_N) ,$$

$$\bar{G}_t = N_t^{-1} \sum_{i=1}^{N_t} \gamma_{ii} , \qquad \underset{\sim}{y}_N = (y_1, y_2, \ldots, y_N) .$$

If the matrix $\underset{\sim}{Z}'_N \underset{\sim}{P}_N^{-1} \underset{\sim}{Z}_N$ is singular it is understood that the dimension should be reduced by removing x_i, or $\gamma_{ii}^{\frac{1}{2}}$, or γ_{ii}, or some combination of these variables from the model.

PROOF. The design variance of the estimator (38) under replacement sampling is

$$n_t^{-1} N_t^{-2} \sum_{i=1}^{N_t} p_i^{-1}[y_i - N_t p_i \bar{Y}_t - C_2(1 - N_t p_i) - C_3(\gamma_{ii} - N_t p_i \bar{G}_t)$$
$$C_4(x_i - N_t p_i \bar{X}_t)]^2 .$$

Minimizing

$$\sum_{i=1}^{N_t} p_i^{-1}[y_i - C_1 N_t p_i - C_2 - C_3 \gamma_{ii} - C_4 x_i]^2 .$$

with respect to (C_1, C_2, C_3, C_4) we obtain $\underset{\sim}{C}_t$. ∎

RESULT 6. Let model (35) hold. In the class of linear conditionally model unbiased predictors of $\bar{Y}_t$, the predictor

$$N_t^{-1}[(N_t - n_t) \bar{\underset{\sim}{Z}}_{N-n,t} \hat{\underset{\sim}{\beta}}_t + n_t \bar{y}_t] = \bar{\underset{\sim}{Z}}_t \hat{\underset{\sim}{\beta}}_t \tag{40}$$

has smallest conditional model variance, where

$$\hat{\underset{\sim}{\beta}}_t = (\underset{\sim}{Z}'_n \underset{\sim}{\Gamma}_n^{-1} \underset{\sim}{Z}_n)^{-1} \underset{\sim}{Z}'_n \underset{\sim}{\Gamma}_n^{-1} \underset{\sim}{y}_n , \tag{41}$$

$$\bar{\underset{\sim}{Z}}_{N-n,t} = (N_t - n_t)^{-1} \sum_{i=1}^{N_t - n_t} \underset{\sim}{z}_i ,$$

$$\underset{\sim}{\Gamma}_n = \text{diag}(\gamma_{11}, \gamma_{22}, \ldots, \gamma_{nn}), \text{ and } \underset{\sim}{y}_n = (y_1, y_2, \ldots, y_n) .$$

PROOF. The left expression in (40) is the standard linear model result. That the left and right expressions are equal follows from the property of the least squares solution;

$$\sum_{i=1}^{n_t} \gamma_{ii}^{-1} (y_i - \underset{\sim}{z}_i \hat{\underset{\sim}{\beta}}) \underset{\sim}{z}_i = \underset{\sim}{0} \tag{42}$$

which implies that

$$\bar{y}_t = \bar{\underset{\sim}{z}}_t \hat{\underset{\sim}{\beta}}_t \ . \tag{43}$$

THEOREM. Let the sequence of populations and samples be as defined. Let model (35) hold. Let the requisite limits exist. Let $\hat{Y}_t = \bar{\underset{\sim}{Z}}_t \hat{\underset{\sim}{\beta}}_t$ and let the inclusion probabilities be

$$\pi_i = n_t \left(\sum_{i=1}^{N_t} \gamma_{ii}^{\frac{1}{2}} \right)^{-1} \gamma_{ii}^{\frac{1}{2}} \ . \tag{44}$$

Then

$$AV(\hat{\bar{Y}}_t - \bar{Y}_t) = M_t \sigma^2 + O(n_t^{-\frac{3}{2}}) \ ,$$

where M_t is defined in Result 4. Furthermore, the design mean square error

$$E\{(\hat{\bar{Y}}_t - \bar{Y})^2\} = E\left\{ \left(\sum_{i=1}^{n_t} (N_t \pi_i)^{-1} e_{it} \right)^2 \right\} + O(n_t^{-\frac{3}{2}}) \ ,$$

where

$$e_{it} = y_i - \underset{\sim}{z}_i \underset{\sim}{C}_t \ ,$$

$$\underset{\sim}{C}_t = (\underset{\sim}{Z}_N' \underset{\sim}{\pi}_N^{-1} \underset{\sim}{Z}_N)^{-1} \underset{\sim}{Z}_N \underset{\sim}{\pi}_N^{-1} \underset{\sim}{y}_N \ ,$$

$$\underset{\sim}{\pi}_N = \operatorname{diag}(\pi_1, \pi_2, \ldots, \pi_N) \ .$$

REFERENCES

Aggarwal, O. P. (1959), "Bayes and Minimax Procedures in Sampling From Finite and Infinite Populations I," Ann. Math. Statist., 30, 206-218.

Allen, E. M. (1960), "Why are Research Grant Applications Disapproved?", Science, 132, 1532-1534.

Basu, D. (1971), "An Essay on the Logical Foundations of Survey Sampling, Part One." In Godambe, V. P. and Sprott, D. A., Eds., Foundations of Statistical Inference, Toronto: Holt, Rinehart and Winston.

Cassel, C., Särndal, C., and Wretman, J. H. (1977), Foundations of Inference in Survey Sampling, New York: John Wiley and Sons.

Cochran, W. G. (1946), "Relative Accuracy of Systematic and Stratified Random Samples for a Certain Class of Populations," Ann. of Math. Statist., 17, 164-177.

Cochran, W. G. (1977), Sampling Techniques, New York: John Wiley and Sons.

Dalenius, T. and Gurney, M. (1951), "The Problem of Optimum Stratification II," Skandinavisk Aktuarietidskript, 133-148.

Deming, W. E. (1950), Some Theory of Sampling, New York: John Wiley and Sons.

Deming, W. E. and Stephan, F. (1941), "On the Interpretation of Censuses as Samples," J. Amer. Statist. Assoc., 36, 45-49.

Durbin, J. (1967), "Design of Multistage Surveys for the Estimation of Sampling Errors," Applied Statistics, 26, 152-164.

Fuller, W. A. (1970), "Sampling With Random Stratum Boundaries," J. Roy. Statist. Soc. B, 32, 203-226.

Fuller, W. A. (1975), "Regression Analysis for Sample Survey," Sankhyā C, 37, 117-132.

Godambe, V. P. (1955), "A Unified Theory of Sampling From Finite Populations," J. Roy. Statist. Soc. B, 17, 269-278

Godambe, V. P. and Joshi, V. M. (1965), "Admissibility and Bayes Estimation in Sampling Finite Populations, 1," Ann. Math. Statist., 36, 1707-1722.

Godambe, V. P. and Sprott, D. A., Eds. (1971), Foundations of Statistical Inference, New York: Holt.

Graham, J. E. and Rao, J.N.K. (1978), "Sample Surveys: Theory and Practice." In Hogg, R. V. Ed., Studies in Statistics MAA Studies in Mathematics Vol. 19. Mathematical Association of America.

Hájek, J. (1959), "Optimum Strategy and Other Problems in Probability Sampling," Casopis Pro Pestovani Matematiky, 84, 387-423.

Hájek, J. (1960), "Limiting Distributions in Simple Random Sampling From a Finite Population," Publ. Math. Inst. Hung. Acad. Sci., 5, 361-374.

Hansen, M. H. and Hurwitz, W. N. (1943), "On the Theory of Sampling From Finite Populations," Ann. Math. Statist., 14, 333-362.

Hansen, M. H. Hurwitz, W. N. (1949), "On the Determination of Optimum Probabilities in Sampling," Ann. Math. Statist., 20, 426-432.

Hansen, M. H., Hurwitz, W. N., and Madow, W. G. (1953), Sample Survey Methods and Theory Vols. I and II, New York: John Wiley and Sons.

Hanurav, T. V. (1968), "Hyper-Admissibility and Optimum Estimator for Sampling Finite Populations," Ann. Math. Statist., 39, 621-642.

Horvitz, D. G. and Thompson, D. J. (1952), "A Generalization of Sampling Without Replacement From a Finite Universe," J. Amer. Statist. Assoc., 47, 663-685.

Isaki, C. T. (1970), "Survey Designs Utilizing Prior Information," Unpublished Ph.D. Thesis, Library, Ames, Iowa: Iowa State University.

Jessen, R. J. (1978), Statistical Survey Techniques, New York: John Wiley and Sons.

Johnson, N. L. and Smith, H., Eds. (1969), New Developments in Survey Sampling, New York: John Wiley and Sons.

Lazarsfeld, P. F. (1935), "The Art of Asking Why in Marketing Research," National Marketing Review, 1, 26-38.

Madow, W. G. (1948), "On the Limiting Distributions of Estimates Based on Samples From Finite Universe," Ann. Math. Statist., 19, 535-545.

Namboodiri, N. K. (1978), Survey Sampling and Measurement, New York: Academic Press.

Neyman, J. (1934), "On the Two Different Aspects of the Representative Method: The Method of Stratified Sampling and the Method of Purposive Selection," (with discussion), J. Roy. Statist. Soc., 97, 558-625.

Neyman, J. (1935), "On the Problem of Confidence Intervals," Ann. Math. Statist., 6, 111-116.

Payne, S. L. (1951), The Art of Asking Questions, Princeton, N.J.: Princeton University Press.

Rao, J.N.K., Hartley, H. O., and Cochran, W. G. (1962), "On a Simple Procedure of Unequal Probability Sampling Without Replacement," J. Roy. Statist. Soc. B., 24, 482-491.

Royall, R. M. (1970), "On Finite Population Sampling Theory Under Certain Linear Regression Models," Biometrika, 57, 377-387.

Sampford, M. R. (1967), "On Sampling Without Replacement With Unequal Probabilities of Selection," Biometrika, 54, 499-513.

Sirken, M. G. (1970), "Household Surveys with Multiplicity," J. Amer. Statist. Assoc., 65, 257-266.

Smith, T.M.F. (1976), "The Foundations of Survey Sampling: A Review," J. Roy. Statist. Soc. A, 139, 183-204.

Snedecor, G. W. and Cochran, W. G. (1967), Statistical Methods, Ames, Iowa: Iowa State University Press.

Sukhatme, P. V. and Sukhatme, B. V. (1970), Sampling Theory of Surveys With Applications, Ames, Iowa: Iowa State University Press.

Williams, B. (1978), A Sampler on Sampling, New York: John Wiley and Sons

Yates, F. (1937), "Applications of the Sampling Technique to Crop Estimation and Forecasting," Transactions of the Manchester Statistical Society Session 1936-1937. Cited in Hansen, Hurwitz, and Madow, Vol. 1, (1953, p. 72).

Yates, F. and Grundy, P. M. (1953), "Selection Without Replacement From Within Strata With Probability Proportional to Size," J. Roy. Statist. Soc. B, 15, 253-261.

Proceedings of Symposia in Applied Mathematics
Volume 23
1980

THE ANALYSIS OF VARIANCE

Peter W. M. John

1. INTRODUCTION. The title, The Analysis of Variance, for this field is almost self explanatory. We take a set of observations, y_{ij}, and let S_y denote the sum of the squares of the deviations of the observations from the grand mean. We then analyse S_y by breaking it up into portions or components that explain the scatter in the data. The analysis of variance is the basis of the design of experiments. It was introduced in England about fifty years ago by R.A. Fisher for agricultural experiments. Its use has spread to experiments in other fields, such as engineering and psychology. Numerous examples of applications to agricultural experiments will be found in the book by Cochran and Cox (1957). A mathematical treatment is given in John (1971), and a short introduction in Hogg (1979).

2. COMPARING TWO TREATMENTS. We begin with a simple comparative experiment. Suppose that we take a field and divide it into $2r$ plots of equal size. We then sow variety A of wheat in r of the plots, and variety B in the other r. We take pains, of course, to choose the r plots that are to contain variety A by some random procedure, so as to minimize the danger that we inadvertently introduce a bias by giving A more than its fair share of the more fertile plots.

Let y_{ij} be the yield on the j-th plot that was sown with the i-th variety; $i=1$ for A, 2 for B. Our model for y_{ij} is

$$y_{ij} = \mu_i + e_{ij}, \qquad j = 1,2,\ldots,r,$$

where μ_i is an unknown constant (the 'true' yield of the i-th variety), and e_{ij} represents random variation due to differences in fertility, cultivation depth of watering and other vagaries of nature. Unfortunately it has become the custom to refer to e_{ij} as errors, which in industrial experimentation has the connotation of mistakes; on the contrary, the e_{ij} terms represent the natural variability (or 'noise') in the system.

1980 Mathematics Subject Classification - 62K

The 'error' terms, e_{ij}, are random variables, and we make the following basic assumptions about their probabilistic behavior:

(i) The e_{ij} are statistically independent,

(ii) each e_{ij} is normally distributed with $E(e_{ij}) = 0$ and $V(e_{ij}) = \sigma^2$. Note that σ^2 is the same for all the plots; this is the assumption of homoscedasticity.

The primary purpose of our experiment is to compare the yields of the two varieties, which involves estimating $\mu_1 - \mu_2$. We estimate μ_i by $\bar{y}_i$, which is the average of the yields of the r plots which contain the i-th variety. This is both the least squares estimate of μ_i and the maximum likelihood estimate. The variance of $\bar{y}_i$ is σ^2/r.

Our estimate of $\mu_1 - \mu_2$ is, thus,

$$d = \bar{y}_1 - \bar{y}_2 .$$

The random variable d is normally distributed with $E(d) = \mu_1 - \mu_2$, $V(d) = 2\sigma^2/r$. We could, therefore, if we knew σ, construct a confidence interval for $\mu_1 - \mu_2$ and say, with 95% confidence, that

$$\bar{y}_1 - \bar{y}_2 - 1.96\sqrt{2\sigma^2/r} \leq \mu_1 - \mu_2 \leq \bar{y}_1 - \bar{y}_2 + 1.96\sqrt{2\sigma^2/r} .$$

We need to estimate σ^2. Let $S_i = \sum_j (y_{ij} - \bar{y}_i)^2$; S_i is the sum of the squares of the deviations of the observations on the i-th variety about their mean, and so $S_i/(r-1)$ is an estimate of σ^2 with $(r-1)$ degrees of freedom. We pool the estimates from the two varieties and, writing $S = S_1 + S_2$ take for our estimate of σ^2

$$s^2 = S/(2r-2) .$$

Under normality S/σ^2 has a χ^2 distribution with $2(r-1)$ d.f., and

$$t = \frac{(\bar{y}_1 - \bar{y}_2) - (\mu_1 - \mu_2)}{s\sqrt{(2/r)}}$$

has Student's t distribution with $2(r-1)$ d.f. The confidence interval becomes

$$y_1 - y_2 - t^* \sqrt{2s^2/r} \leq \mu_1 - \mu_2 \leq y_1 - y_2 + t^* \sqrt{2s^2/r} ,$$

where t^* is the critical value of t for $2(r-1)$ d.f. If we are interested in testing the hypothesis H_0: $\mu_1 = \mu_2$, we ask whether the interval contains zero; if so we accept H_0; if not we reject H_0. Equivalently we declare that $\mu_1 \neq \mu_2$ if, and only if,

$$|\bar{y}_1 - \bar{y}_2| > t^* \sqrt{2s^2/r} .$$

The quantity $t^* \sqrt{2s^2/r}$ is called Fisher's LSD (least significant difference).

3. THE ONE-WAY ANALYSIS OF VARIANCE. We now generalize in two ways. First we shall change our vocabulary to speak of applying 'treatments' to 'plots'. The 'plots' are basic experimental units; they may be actual plots of ground, or runs in a pilot plant, or pieces of rubber, or young pigs. The 'treatments' may be different varieties of wheat, different temperatures of operation, different methods of vulcanizing, or different rations. The 'yields' may be bushels per acre, percentages of conversion of crude oil, tensile strengths or weight gains.

We now consider $n = rt$ plots; t treatments are applied, one to each plot, r plots to each treatment. Since there is only one classification of the data, namely by treatments, this is the one-way analysis of variance.

Our model could be the same as the model in the previous section

$$y_{ij} = \mu_i + e_{ij} ;$$

where j is now the yield on the j-th plot with the i-th treatment. It is customary to rewrite the model as

$$y_{ij} = \mu + \tau_i + e_{ij} .$$

Here μ represents a grand mean and τ_i the 'effect' of the i-th treatment; there is a side condition $\Sigma\tau_i = 0$ (without a side condition the model would be overparametrized). We write $\underset{\sim}{\tau} = (\tau_1, \tau_2, \ldots, \tau_t)'$. As before the 'error' terms e_{ij} are independent and normal with $E(e) = 0$, $V(e) = \sigma^2$.

The parameters μ, τ_i are assumed to be unknown constants. We are primarily interested in comparing treatments and we ask two questions: first, do the treatments really differ in respect to yield and, secondly, if the answer to the first question is yes, which treatments differ?

The first question is equivalent to the hypothesis H_0: $\underset{\sim}{\tau} = \underset{\sim}{0}$. The second involves $\binom{t}{2}$ hypotheses of the form H_0: $\tau_h = \tau_i$.

The parameters are estimated by the method of least squares in the following way:

$$\hat{\mu} = \bar{y} ,$$

the average of all n observations; then

$$\hat{\mu} + \hat{\tau}_i = \bar{y}_i ,$$

so that

$$\hat{\tau}_i = \bar{y}_i - \bar{y} .$$

These estimates are also the maximum likelihood estimates. The variance is estimated by a pooled estimate from all t treatments. The difference $d_{ij} = y_{ij} - \bar{y}_i$ is called the (ij)-th residual. The expression

$$S_e = \sum_i \sum_j d_{ij}^2$$

is called the residual sum of squares; σ^2 is estimated by $s^2 = S_e/[t(r-1)]$; S_e/σ^2 has a χ^2 distribution with $t(r-1)$ d.f.

Let $S_y = \sum\sum_{ij}(y_{ij}-\bar{y})^2$, the sum of the squares of the observations about their common mean. This is usually called the total sum of squares. If the treatments are equal with respect to yield, i.e. if H_0: $\tau_1 = \tau_2 = \ldots$ is true, then $E(S_y) = (n-1)\sigma^2$ and S_y/σ^2 has a χ^2 distribution with $(n-1)$ d.f. We now break Sy into two components using an algebraic identity:

$$\sum\sum_{ij}(y_{ij}-\bar{y})^2 = r\sum_i(\bar{y}_i-\bar{y})^2 + \sum\sum_{ij}(y_{ij}-\bar{y}_i)^2 .$$

$$S_y = S_t + S_e .$$

The component S_t is called the sum of squares for treatments; it represents the scatter among the treatment means. Whether H_0 is true or not, $E(S_e) = t(r-1)\sigma^2$. If H_0 is true, $E(S_t) = [(n-1)-t(r-1)]\sigma^2 = (t-1)\sigma^2$, and S_t/σ^2 has a χ^2 distribution with $(t-1)\sigma^2$. In that case the corresponding mean square

$$M_t = S_t/(t-1)$$

should be an unbiased estimate of σ^2. If H_0 is false, M_t will tend to be larger than σ^2 because of the treatment differences. We may therefore test H_0 by looking at the ratio $\mathcal{F} = M_t/s^2$ and rejecting H_0 if $\mathcal{F}$ is appreciably greater than unity.

More formally, under H_0, $\mathcal{F}$ has the $F[t-1,t(r-1)]$ distribution and we reject H_0, if and only if,

$$\mathcal{F} > F^*[(t-1),t(r-1)]$$

where F^* is the upper-tail critical value of F.

We summarize this part of the discussion in the Analysis of Variance Table.

Source	S.S.	d.f.	M.S.	F
Total	S_y	n-1		
Treatments	S_t	t-1	M_t	M_t/s^2
Residual	S_e	t(r-1)	s^2	

From a computational standpoint it is easier to write $T_i = \sum_j y_{ij}$, $G = \sum_{ij}\sum y_{ij}$, and $C = G^2/n$. Then

$$S_y = \sum_{ij}\sum y_{ij}^2 - C \ , \quad S_t = \sum_i T_i^2/r - C \ , \quad S_e = S_y - S_t \ .$$

An alternative approach is to set up the generalized likelihood ratio test for the hypothesis H_0: $\underset{\sim}{\tau} = 0$ against the alternative H_A: $\underset{\sim}{\tau} \neq \underset{\sim}{0}$. This will also lead us to the F-test that has just been presented.

The necessity for imposing a side condition on the parameters is apparent when we note that, if we were to add any nonzero number δ to μ, and subtract δ from each τ_i, we should have the same expected values for y_{ij}. The method of least squares calls for choosing the estimates $\hat{\mu}$, $\hat{\tau}_i$ so as to minimize $S_e = \sum_{ij}\sum (y_{ij} - \hat{\mu} - \hat{\tau}_i)^2$. This leads to $t+1$ normal equations. The first, given by $\partial S_e/\partial\hat{\mu} = 0$, is

$$\sum_{ij}\sum y_{ij} = n\hat{\mu} + r\sum_i \hat{\tau}_i \ .$$

There are t other equations, corresponding to $\partial S_e/\partial\hat{\tau}_i$,

$$\sum_j y_{ij} = r\hat{\mu} + r\hat{\tau}_i \ .$$

There are $t+1$ equations in $t+1$ unknowns, but the system is singular because the sum of the last t equations is equal to the first equation. The side condition $\sum \tau_i = 0$ removes the singularity. It can be shown that any linear constraint $\underset{\sim}{c}'\underset{\sim}{\tau} = 0$, where $\underset{\sim}{c}'\underset{\sim}{1} \neq 0$ will suffice to provide a generalized inverse for the normal equations, see John (1971).

4. COMPARING TREATMENTS. Having rejected H_0: $\underset{\sim}{\tau} = \underset{\sim}{0}$, we turn now to the question of which treatments differ. We estimate $\mu_h - \mu_i$ by $\bar{y}_h - \bar{y}_i$ and argue, as in section 2, that $\bar{y}_h - \bar{y}_i$ is normally distributed with variance

$2\sigma^2/r$. An obvious procedure, therefore, is again to compute the least significant difference

$$LSD = t^* \sqrt{2s^2/r} ,$$

where t^* is the critical value of t corresponding to $t(r-1)$ d.f. and to our choice of α. Then we apply the LSD to each pair of means and declare that $\mu_h \neq \mu_i$ if and only if $|\bar{y}_h - \bar{y}_i|$ exceeds the LSD.

There is an objection to this procedure; it leads to too many differences being declared significant. The rationale for the objection is this. If the treatments are equivalent, the means $\bar{y}_i$ may be regarded as a random sample from a normal population with variance $v^2 = 2\sigma^2/r$. If there were only two treatments it would be correct to say that $p(|\bar{y}_1 - \bar{y}_2| > 1.96v) = 0.05$. However if $t > 2$, the probability that the largest and smallest of the treatment means differ by more than $1.96v$ is going to be more than five per cent. Tukey has suggested that we should be looking at the range of a sample of t normal variables.

This changes the emphasis of the error rate. When we developed Fisher's LSD we were thinking of a risk α (e.g. 0.05) of making a type I error in each contrast, $\bar{y}_h - \bar{y}_i$, that we tested. When we switch to Tukey's approach we regard α as the risk that, amongst $t(t-1)/2$ comparisons, one will be found significant when $\tau_1 = \tau_2 = \ldots = 0$. This represents an error risk of α per experiment.

Tukey replaces the LSD by his HSD (honestly significant difference), which is defined by

$$HSD = q(\alpha, t, \phi) \sqrt{s^2/r} ,$$

with $\phi = t(r-1)$. In this expression $q(\alpha, t, \phi)$ is the critical value of the Studentized range, corresponding to a risk probability α, for a sample of size t and an estimate of the variance based on ϕ d.f.

The Studentized range is the range (i.e. $x_{max} - x_{min}$) of a sample from a normal population, divided by an estimate of the standard deviation which is based upon a chi-squared statistic. It should be noted that, if t is increased by adding more treatments to the experiment, the HSD is automatically increased.

It is argued that the HSD swings the pendulum too far in the other direction; it is too conservative. One way to compensate is to take $\alpha = 0.10$ rather than 0.05. However some statisticians prefer to use one or other of the two compromises that will be presented shortly. They are the Newman-Keuls test and Duncan's Multiple Range test.

We shall illustrate all four tests by an example with $t = 6$, $r = 5$, $\phi = 24$. The treatment means have been labelled A through F in ascending order. The computation of the various significant differences and the conclusions drawn by using the different methods are shown in Table I.

In the Newman-Keuls and Duncan tests we begin, as we have in our example, by arranging the treatment means in ascending order. In our example $\bar{y}_F - \bar{y}_A$ is the range of $p = 6$ means; $\bar{y}_E - \bar{y}_B$ is the range of $p = 4$ means. The critical value to be used in each pair depends upon the value of p.

The Newman-Keuls difference, NKD, is defined as

$$\mathrm{NKD} = q(\alpha, p, \phi)\sqrt{(s^2/r)}\ .$$

Duncan's significant difference, which we shall write as DSD, is defined by

$$\mathrm{DSD} = q(\alpha_p\,, p, \phi)\sqrt{(s^2/r)}\ ,$$

where $\alpha_p = 1 - (1-\alpha)^{p-1}$. Special tables of α_p are available.

There are numerous other multiple comparison tests. The four that we have mentioned are the most commonly used. Chew's monograph (1977) covers the problem in more detail than time permits here, and he gives tables of the critical values of the relevant statistics, $q(\alpha, p, t)$ and $q(\alpha_p\,, p, t)$.

TABLE I

Significant Differences Between Treatments

$t = 6$, $r = 5$, $\phi = 24$, $s^2 = 33.78$

Treatment Means

A	B	C	D	E	F
39.3	45.2	48.4	50.4	55.5	58.2

with $\alpha = 0.05$, $t^* = 2.064$, $q(0.05, 6, 24) = 4.373$

$\sqrt{(s^2/r)} = 2.599$, $\sqrt{(2s^2/r)} = 3.676$

LSD = 7.59 HSD = 11.37

For The Newman-Keuls and Duncan Tests

p =	2	3	4	5	6
q(0.05,p,24)	2.919	2.532	3.901	4.166	4.373
NKD	7.59	9.18	10.14	10.83	11.37
$q(\alpha_p, p, 24)$	2.919	3.066	3.160	3.226	3.276
DDD	7.59	7.97	8.21	8.38	8.51

Conclusions

LSD	HSD	N-K	Duncan
F > A,B,C,D	A,B	A,B	A,B,C
E > A,B	A	A,B	A,B
D > A		A	A
C > A			A

We have only considered simple comparisons, $\bar{y}_h - \bar{y}_i$, between the treatment means, estimating the corresponding comparison $\mu_h - \mu_i$. More generally, let $\underset{\sim}{c}$ be a vector of t elements such that $\underset{\sim}{c}'\underset{\sim}{1} = 0$; and let $\underset{\sim}{\tau}'$ be the vector $(\hat{\tau}_1, \hat{\tau}_2, \ldots, \hat{\tau}_t)'$. The set of contrasts $\underset{\sim}{c}'\underset{\sim}{\tau}$ is the set of estimable functions of the treatment effects. The corresponding estimates are $\underset{\sim}{c}'\hat{\underset{\sim}{\tau}} = \Sigma c_i \bar{y}_i$. They form a vector space, which has rank $(t-1)$. This vector space may be considered as a subspace of the larger vector space of all contrasts in the observations and we may regard the larger space as being partitioned into two orthogonal subspaces: the estimation and error subspaces.

5. THE COMPONENTS OF VARIANCE MODEL. In the previous sections we have taken the effects, τ_i, to be unknown constants. In such areas as sampling and genetics a random effects model is sometimes appropriate. Suppose that we take q letters at random, and then take r siblings from each litter, observing for each animal some such quantity as its weight when it is ten days old.

We let y_{ij} be the weight of the j-th animal in the i-th litter, and write the model

$$y_{ij} = \mu + a_i + e_{ij} .$$

Here a_i is not a constant but a random variable, assumed to be normally distributed with zero expectation and variance σ_A^2. We are interested in estimating σ_A^2, the variance between litters. We assume that the a_i and e_{ij} are all independent of one another; $V(e_{ij})$ is now called the variance within litters.

It can be shown that $\sum_{ij}\sum(y_{ij}-\bar{y}_i)^2$, which we now call the sum of squares between litters and denote by S_a, has for its expectation

$$E(S_a) = (q-1)\sigma^2 + r(q-1)\sigma_A^2$$

and so we may test the hypothesis $H_0: \sigma_A^2 = 0$ by the F test of section 3, and estimate σ_A^2 by

$$\hat{\sigma}_A^2 = [S_a-(q-1)s^2]/r(q-1) .$$

6. THE RANDOMIZED BLOCK EXPERIMENT. The precision of our comparisons between treatments depends upon the size of the error variance. If we can reduce the variance we thereby improve the precision. In an agronomy experiment we may divide the field into strips called blocks. The plots in each block are more homogeneous than plots in different blocks. We could take one block on low land, one on high land, and so on. Each block is divided into t plots, and in each block one plot is assigned to each treatment by a randomisation procedure. In an industrial experiment the blocks might be different machines for carrying out the same operation, different batches of raw material or different technicians operating the equipment.

We modify our model in the following way. Let y_{ij} be the yield on the plot in the j-th block which receives the i-th treatment; then

$$y_{ij} = \mu + \tau_i + \beta_j + e_{ij} ,$$

where β_j represents the effect of the j-th block.

In section 3 we separated out from S_y, a sum of squares, S_t, for the difference between treatments. We now denote the treatment means by $\bar{y}_{i.}$, and the block means by $\bar{y}_{.j}$. We separate an additional component

$$S_b = \sum_{ij}\sum(y_{ij}-\bar{y}_{.j})^2 ,$$

the sum of squares between blocks. The residual sum of squares becomes

$$S_e = S_y - S_t - S_b ,$$

with expected value $(n-t-b+1)\sigma^2$, so that the estimate of σ^2 is $s^2 = S_e/\phi$, where $\phi = (n-t-b+1)$. The estimate s^2 has ϕ d.f.

We had imposed a side condition $\Sigma\tau_i = 0$ upon the treatment effects; we impose a similar side condition $\Sigma\beta_j = 0$ upon the block effects. The least squares estimate of β_j is

$$\hat{\beta}_j = \bar{y}_{.j} - \bar{y} .$$

An important facet of this experimental design is that each treatment appears exactly once in each block. The blocks are said to be orthogonal to the treatments. We now have two orthogonal vector subspaces; they are orthogonal in the sense that any estimating contrast in the treatment subspace is orthogonal to any estimating contrast in the block subspace. What this means in practice is that estimates of treatment comparisons are independent of block differences. The various methods of obtaining significant differences between treatment means are still used; the only change is the change in the value of ϕ.

7. FACTORIAL EXPERIMENTS. Consider an industrial experiment in a chemical plant in which we are able to vary the temperature and the pressure, and suppose that we choose to make runs at t different temperatures and at p different pressures. A completely balanced experiment would call for making $n = pt$ runs in the plant, one at each combination of temperature and pressure. This is a simple factorial experiment with two factors: temperature at t levels and pressure at p levels.

More generally we envisage an experiment with two factors, A and B, with a and b levels respectively. Let y_{ij} denote the yield of the run made at the i-th level of A and the j-th level of B. We modify the model of the previous section only slightly by writing

$$y_{ij} = \mu + \alpha_i + \beta_j + e_{ij} ;$$

α_i is the effect of the i-th level of A, β_j is the effect of the j-th level of B, with $\Sigma\alpha_i = 0$, $\Sigma\beta_j = 0$; we make the usual assumptions about the distribution of the e_{ij} terms. The basic analysis follows the analysis of the randomized block design closely. The orthogonality now means that any estimating contrast between the means at the various levels of A is independent of every contrast between the levels of B and vice versa.

We conclude by giving the analysis of variance table for this experiment.

Table II

Analysis of Variance Table for a Two-factor Experiment

Source	Sum of Squares	d.f.
Total	$\sum_{ij}\sum y_{ij}^2 - C$	ab - 1
A	$S_A = \sum_{ij}\sum (y_{ij} - \bar{y}_{i\cdot})^2 = \sum_i T_{i\cdot}^2/b - C$	a - 1
B	$S_B = \sum_{ij}\sum (y_{ij} - \bar{y}_{\cdot j})^2 = \sum_j T_{\cdot j}^2/a - C$	b - 1
Residual	$S_e = S_y - S_A - S_B$	(a-1)(b-1)

$$T_{i\cdot} = \sum_j y_{ij}, \qquad T_{\cdot j} = \sum_i y_{ij}$$

BIBLIOGRAPHY

1. V. Chew, "Comparisons among treatment means in an analysis of variance", Publication ARS/H/6 Agricultural Research Service, USDA, Washington, D.C., 1977. (This may be obtained by writing to Mr. Chew at the University of Florida, Room 217, Rolfs Hall, Gainesville, Fla. 32611.)

2. W.G. Cochran and G.M. Cox, Experimental Designs, John Wiley, New York, 1957, 2nd Edition, 1-116.

3. R.V. Hogg, "Studies in statistics", Mathematical Association of America, Studies in Mathematics, 19(1978), 8-38.

4. P.W.M. John, Statistical Design and Analysis of Experiments, Macmillan, New York, 1971, 39-79.

DEPARTMENT OF MATHEMATICS
THE UNIVERSITY OF TEXAS AT AUSTIN
AUSTIN, TEXAS 78712

Proceedings of Symposia in Applied Mathematics
Volume 23
1980

NONPARAMETRIC STATISTICAL TESTS OF HYPOTHESES

Ronald H. Randles

1. DISTRIBUTION-FREE TESTS. Methods of statistical inference all make certain assumptions about the underlying distribution of the observations. This presentation will concentrate on statistical tests of hypotheses, and we will describe certain techniques for creating robustness with respect to the distributional assumptions made in these tests. For example, if $X_1, \cdots, X_n$ denotes a random sample from a normal distribution with unknown mean μ and unknown variance σ^2, then to test $H_0 : \mu = 0$ versus $H_1 : \mu > 0$, the one-sample t-test rejects H_0 in favor of H_1 whenever

$$(1.1) \qquad T^+ = \frac{\overline{X} - 0}{S/\sqrt{n}} > t^+_{\alpha;n-1},$$

where $\overline{X}$ is the sample mean, S^2 is the sample variance with an $n-1$ in the denominator, and $t_{\alpha;n-1}$ is the upper $\alpha^{\underline{th}}$ percentile of a t-distribution with $n-1$ degrees of freedom. The ability of this test procedure to maintain the prescribed significance level α and to detect μ-values in H_1 is highly dependent on the normal distribution assumptions. What we need is a structure that will enable us to control the probability of making a type I error with minimal assumptions made about the underlying distributions.

DEFINITION 1.2. A statistic T (possibly vector valued) is said to be distribution-free over a class of underlying distributions $\mathcal{A}$, if the distribution of T is the same for every distribution model in $\mathcal{A}$. □

If a testing procedure can be constructed using a statistic which has the same null hypothesis distribution over a wide class of distributional assumptions $\mathcal{A}$, then this distribution-free property will protect the probability of making a type I error over a broad set of assumptions. It is hoped that the class of distributions $\mathcal{A}$ might include many different common distribution families, such as uniform, normal, double exponential, gamma, etc.

This presentation will emphasize the use of ranks, one of the prominent methods for achieving a distribution-free property. To be specific, let

---

1980 Mathematics Subject Classification 68G10.

$Z_1,\cdots,Z_N$ denote a random sample of size N from some continuous population. The utility of ranks is built on the following foundation:

INTUITIVE PROPERTY 1.3. Any ordering among the observations $Z_1,\cdots,Z_N$ is just as likely as any other. □

Ranks provide a convenient expression of this property.

DEFINITION 1.4. We assign the smallest observation the <u>rank</u> 1, the second smallest gets <u>rank</u> 2, etc., and the largest gets <u>rank</u> N. Let R_i^* denote the rank of Z_i among $Z_1,\cdots,Z_N$. □

For example, consider

$$z_1 = 4.8 \quad z_2 = 3.5 \quad z_3 = 1.6 \quad z_4 = 7.8 \quad z_5 = 2.0$$
$$r_1^* = 4 \quad r_2^* = 3 \quad r_3^* = 1 \quad r_4^* = 5 \quad r_5^* = 2.$$

The intuitive property 1.3 then states that each of the $N!$ possible rankings is just as likely as any other; hence they each have probability $1/N!$. A formal statement of this result (which might be termed the "fundamental theorem of rank tests") is as follows.

THEOREM 1.5. Consider $\underset{\sim}{Z} = (Z_1,\cdots,Z_N)$ and the corresponding rank vector $\underset{\sim}{R} = (R_1^*,\cdots,R_N^*)$. If $Z_1,\cdots,Z_N$ is a random sample of size N from any continuous population, then for any $\underset{\sim}{r} \in \mathcal{R}$,

$$P[\underset{\sim}{R}^* = \underset{\sim}{r}] = \frac{1}{N!}$$

where $\mathcal{R} = \{\text{all } N! \text{ permutations of the integers } (1,\cdots,N)\}$.

PROOF. Select $\underset{\sim}{r} \in \mathcal{R}$ and define a corresponding vector $\underset{\sim}{d}$ such that d_i is the position of the integer i within $\underset{\sim}{r}$. Since $(Z_{d_1},\cdots,Z_{d_N})$ has the same joint distribution as $(Z_1,\cdots,Z_N)$, it follows that

$$\begin{aligned} P[\underset{\sim}{R}^* = \underset{\sim}{r}] &= P[Z_{d_1} < Z_{d_2} < \cdots < Z_{d_N}] \\ &= P[Z_1 < Z_2 < \cdots < Z_N] \\ &= P[\underset{\sim}{R}^* = (1,2,\cdots,N)]. \end{aligned}$$

Thus every permutation $\underset{\sim}{r}$ in $\mathcal{R}$ is equally likely and the conclusion follows. □

This theorem shows that the rank vector has the same distribution over a very broad class of distributional models. Thus any rank statistic will have a distribution-free property over this same class.

2. TWO-SAMPLE LOCATION PROBLEM.

The two-sample location problem is one of the basic settings of statistical inference and it is also one in which the role and use of ranks is very natural. Suppose we have two independent random samples: $X_1,\cdots,X_m$ from a continuous population with distribution function $F(x)$ and $Y_1,\cdots,Y_n$ from a continuous population with distribution function $F(x-\Delta)$. The quantity Δ is a location (shift) parameter and, when $\Delta > 0$, the Y population is located Δ units higher on the number scale than the X population. We wish to test

$$H_0 : \Delta = 0 \quad \text{versus} \quad H_1 : \Delta > 0.$$

Thus under H_0 the two populations are located in the same place; but under H_1 the Y's tend to be larger than the X's.

We see that under H_0, the observations $(X_1,\cdots,X_m,Y_1,\cdots,Y_n)$ are like a random sample of size $N = m+n$ from some continuous distribution. Let $(Q_1,\cdots,Q_m,R_1,\cdots,R_n)$ denote the corresponding vector of ranks, that is, R_j (Q_i) represents the rank of Y_j (X_i) among all $m+n$ observations. Theorem 1.5 shows that under H_0, this rank vector is uniformly distributed over the $(m+n)!$ permutations of the first $m+n$ positive integers. Moreover, this result holds for any underlying continuous distribution $F(\cdot)$. So any test based on this vector of ranks will be distribution-free over this broad class.

A test statistic for this problem ought to indicate when the alternative is true. Under H_1 the Y's will tend to be larger than the X's and this difference should be reflected in the ranks, the Y's tending to receive larger ranks. Thus one natural rank test statistic rejects H_0 if the sum of the ranks associated with the Y's is large, that is, if

$$S = \sum_{j=1}^{n} R_j \geq d_{\alpha;m,n}. \tag{2.1}$$

This is known as the Mann-Whitney-Wilcoxon test. The values for $d_{\alpha;m,n}$ are determined by constructing the null hypothesis distribution for S based on the equally likely rank vectors described in Theorem 1.5. Since S has a discrete distribution there are only a finite number of natural α-levels associated with this test. Note also that the designated α-level holds over the class of all continuous population distributions.

3. ONE-SAMPLE LOCATION PROBLEM. The one-sample location problem provides a setting in which the rank technique and an interesting variation on that technique are used to create a distribution-free property. Suppose that we have a random sample $X_1,\cdots,X_n$ from a continuous population that is symmetric about θ. Let $F(x-\theta)$ denote the distribution function of the population. Because the distribution is symmetric around θ, it satisfies

$$F(x-\theta) = 1 - F(\theta-x), \quad -\infty < x < \infty. \tag{3.1}$$

Since θ represents the location of the distribution, we test hypotheses about θ, such as

$$H_0 : \theta = 0 \quad \text{versus} \quad H_1 : \theta > 0.$$

Under this H_0 the population is symmetric about zero and under H_1 it tends to produce values greater than zero. Define

$$\Psi_i = \Psi(X_i)$$

where $\Psi(t) = 1,0$ as $t >, \leq 0$. The following lemma provides the key to producing a distribution-free property.

LEMMA 3.2. When $H_0 : \theta = 0$ holds, Ψ_i and $|X_i|$ are stochastically independent.

PROOF. For $t \geq 0$

$$\begin{aligned} P[\Psi_i = 1, |X_i| \leq t] &= P[0 < X_i \leq t] \\ &= F(t) - F(0) \\ &= \tfrac{1}{2}[F(t) - F(0) + F(0) - F(-t)] \quad \text{since } X_i \\ &\quad \text{is continuous and symmetric about } 0, \\ &= P[\Psi_i = 1]P[|X_i| \leq t]. \end{aligned}$$

Similarly,

$$\begin{aligned} P[\Psi_i = 0, |X_i| \leq t] &= P[|X_i| \leq t] - P[\Psi_i = 1, |X_i| \leq t] \\ &= (1 - P[\Psi_i = 1])P[|X_i| \leq t] \quad \text{from above} \\ &= P[\Psi_i = 0]P[|X_i| \leq t]. \end{aligned}$$

Hence Ψ_i and $|X_i|$ are independent. □

The following theorem shows how this result is used to produce a distribution-free property. In it we let R_i^+ denote the rank of $|X_i|$ among $|X_1|, \cdots, |X_n|$.

THEOREM 3.3. Under H_0, the vectors $\underset{\sim}{\Psi} = (\Psi_1, \cdots, \Psi_n)$ and $\underset{\sim}{R}^+ = (R_1^+, \cdots, R_n^+)$ are mutually independent. Moreover, $\underset{\sim}{R}^+$ is uniformly distributed over the $n!$ permutations of the integers $1, \cdots, n$ and $\underset{\sim}{\Psi}$ consists of n independent Bernoulli random variables, each with $p = 1/2$.

PROOF. The mutual independence of $X_1, \cdots, X_n$ together with Lemma 3.2 shows that $\Psi_1, \cdots, \Psi_n, |X_1|, \cdots, |X_n|$ are mutually independent. The Bernoulli nature of each Ψ_i is easy to see. The uniform distribution of $\underset{\sim}{R}^+$ follows from Theorem 1.5 because $|X_1|, \cdots, |X_n|$ are a random sample from a continuous population. □

The conclusion of this theorem holds for any continuous, symmetric distribution $F(\cdot)$. Thus any test statistic based on $\underset{\sim}{\Psi}$ and $\underset{\sim}{R}^+$ will have a distribution-free property for this testing problem.

In this one-sample location problem we want to reject H_0 when the data indicates the population is producing a preponderance of positive values; that is, whenever the positive observations are many and large in absolute value. One test statistic which measures this trait is the <u>Wilcoxon signed-rank test</u> which rejects H_0 in favor of H_1 if the sum of the R_i^+ ranks associated with positive X_i's is large. Thus it rejects H_0 in favor of H_1, if

$$S^+ = \sum_{i=1}^{n} \Psi_i R_i^+ > d^+_{\alpha;n}. \tag{3.4}$$

The critical values $d^+_{\alpha;n}$ are determined by constructing the null distribution of S^+ using Theorem 3.3. The statistic S^+ also has a discrete null hypothesis distribution; hence only selected significance levels (α) occur naturally. Because it is a function of only $\underset{\sim}{\Psi}$ and $\underset{\sim}{R}^+$, the test has a distribution-free property over the entire class of continuous, symmetric distributions.

4. A REGRESSION PROBLEM. Consider a slightly more complicated setting in which a controlled (nonrandom) variable, denoted by c, might potentially be useful in predicting a second variable, X, which is random. The utility of c as a predictor of X is assessed by observing a sample of n pairs of variable values: $(c_1,X_1),\cdots,(c_n,X_n)$. The structure between these two variables is then modeled by a linear regression equation:

$$X_i = \alpha + \beta c_i + E_i \tag{4.1}$$

where the errors $E_1,\cdots,E_n$ are assumed to be a random sample from some continuous population with distribution function $F(x)$. To test whether c is useful as a predictor of X within this structure, we test

$$H_0 : \beta = 0 \qquad \text{versus} \qquad H_1 : \beta \neq 0.$$

To achieve a distribution-free property in this setting, we base our test on the ranks $\underset{\sim}{R} = (R_1,\cdots,R_n)$ where R_i denotes the rank of X_i among $X_1,\cdots,X_n$. Note that under H_0, the variables $X_1 = \alpha + E_1,\cdots,X_n = \alpha + E_n$ are a random sample from a continuous population, and hence by Theorem 1.5, $\underset{\sim}{R}$ is uniformly distributed over the $n!$ permutations of the integers $1,\cdots,n$. Moreover, this distributional result holds for every $F(\cdot)$ of this type, and any test based on these ranks will have a distribution-free property.

For H_1, we need to detect correlation among the c_i's and their observed X_i's or alternatively among the c_i's and their ranks. Computing the usual product-moment correlation coefficient between the c_i's and R_i's yields

$$\hat{\rho} = \frac{\frac{1}{n}\sum_{i=1}^{n} c_i R_i - \bar{c}\bar{R}}{\sqrt{\left[\frac{1}{n}\sum_{i=1}^{n} c_i^2 - \bar{c}^2\right]\left[\frac{1}{n}\sum_{j=1}^{n} R_j^2 - \bar{R}^2\right]}}.$$

But $\bar{R} = \frac{1}{n}\sum_{i=1}^{n} R_i = \frac{1}{n}\sum_{j=1}^{n} j = \frac{n+1}{2}$ and $\sum_{i=1}^{n} R_i^2 = \sum_{j=1}^{n} j^2 = \frac{n(n+1)(2n+1)}{6}$. Therefore, the only random part of this statistic that relates c to X is the term

(4.2) $$S^* = \sum_{i=1}^{n} c_i R_i ;$$

and we would reject H_0 in favor of H_1, if S^* is too large (indicating $\beta > 0$) or too small (indicating $\beta < 0$). This decision is made relative to the discrete null hypothesis distribution of S^* which is found by utilizing Theorem 1.5. The resulting test is distribution-free over the very broad class of all continuous population models for $F(\cdot)$.

5. LARGE SAMPLE NULL HYPOTHESIS DISTRIBUTIONS. When the sample size is large the null hypothesis distribution of each of the above test statistics can be approximated with a standard normal distribution. To illustrate this point consider the one-sample location problem and Wilcoxon signed-rank test statistic

$$S^+ = \sum_{i=1}^{n} \Psi_i R_i^+ .$$

The following result provides the key to developing null hypothesis distributional properties for S^+.

THEOREM 5.1. Under H_0, S^+ has the same distribution as

$$S_0 = \sum_{j=1}^{n} j \Psi_j .$$

PROOF. See, for example, Randles and Wolfe (1979), page 324. □

Using this result we fairly easily find the null hypothesis mean and variance for S^+:

$$E_{H_0}[S^+] = E_{H_0}[S_0] = \sum_{j=1}^{n} j E_{H_0}[\Psi_j]$$

$$= \frac{1}{2} \sum_{j=1}^{n} j = \frac{n(n+1)}{4},$$

and

$$\mathrm{Var}_{H_0}[S^+] = \mathrm{Var}_{H_0}[S_0] = \sum_{j=1}^{n} j^2 \mathrm{Var}_{H_0}[\Psi_j]$$

$$= \frac{1}{4} \sum_{j=1}^{n} j^2 = \frac{n(n+1)(2n+1)}{24} .$$

In addition, we see that S_0 is a weighted sum of some independent Bernoulli random variables. This simple structure makes it relatively easy to apply Liapounov's form of the central limit theorem to show that

$$\frac{S^+ - \frac{n(n+1)}{4}}{\sqrt{\frac{n(n+1)(2n+1)}{24}}}$$

has a limiting standard normal distribution under H_0. Thus, when n is relatively large, the rule would be to reject H_0, if

$$S^+ > \frac{n(n+1)}{4} + z_\alpha \sqrt{\frac{n(n+1)(2n+1)}{24}},$$

where z_α is the upper $\alpha^{\underline{th}}$ percentile of a standard normal distribution.

6. ASYMPTOTIC RELATIVE EFFICIENCY. It is only natural to inquire which of several possible test choices is best; and how much better is it? In a spirit of fair play, we would only seek to compare two tests which are conducted at the same level of significance α. Thus their control of the probability of a type I error is the same; so they must be compared on the basis of their ability to detect alternatives. As in the previous section, for simplicity's sake we will illustrate the comparison of tests in the one-sample location problem in which $X_1,\cdots,X_n$ is a random sample from a continuous population that is symmetric about θ. Suppose we test $H_0 : \theta = 0$ versus $H_1 : \theta > 0$ by rejecting H_0 for large values of some test statistic V; that is, we reject H_0 when $V \geq d$. If $\mu_n(\theta)$ $[\sigma_n(\theta)]$ denotes the mean [standard deviation] of the statistic V when θ is the true parameter value, then Figure 6.1 illustrates the sampling distribution of V under both $H_0 : \theta = 0$ and at $\theta_1 > 0$ a particular value in the alternative H_1. The shaded region in this figure represents the probability of detecting this particular value $\theta_1 > 0$ in the

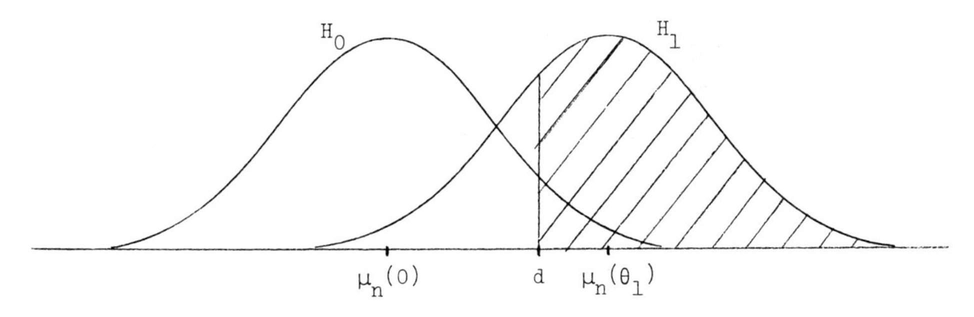

Figure 6.1.

alternative. This probability is called the <u>power</u> of the test at θ_1. Clearly the power of the test varies as θ_1 varies and decreases as θ_1 decreases toward the null hypothesis θ-value of zero.

We would like to compare two tests on the basis of their power, but we must recognize that the power of a test depends on the sample size (n), the significance level of the test (α), the underlying distribution $(F(\cdot))$ and the θ-value in the alternative (θ_1). Thus a concise comparison is difficult. In order to yield a more unified presentation, we therefore make a large sample comparison of the tests. When n is large, the sampling behavior of the test statistic is approximately normal under H_0 and H_1. By examining Figure 6.1, we see that the test with largest power is the one with a mean $\mu_n(\theta_1)$ at θ_1

which has moved furthest away from its mean $\mu_n(0)$ under H_0. Of course, the difference in these means must be measured in terms of the standard deviation of the natural variation of the test statistic. Also the basis of comparison is chosen to be the θ-values in the alternative which are the most difficult to detect, namely the θ-values closest to the null hypothesis $\theta = 0$. Thus we study

$$\mu_n'(0)/\sigma_n(0),$$

the change in the mean function at $\theta = 0$, measured in standard deviations.

DEFINITION 6.2. The <u>asymptotic relative efficiency</u> of the test based on statistic S with mean function $\mu_{n,S}(\theta)$ and st. deviation $\sigma_{n,S}(\theta)$ relative to the test based on statistic T with mean function $\mu_{n,T}(\theta)$ and standard deviation $\sigma_{n,T}(\theta)$ is

$$\text{ARE}(S,T) = \lim_{n\to\infty}\left[\frac{\mu_{n,S}'(0)}{\sigma_{n,S}(0)}\right]^2 \Big/ \left[\frac{\mu_{n,T}'(0)}{\sigma_{n,T}(0)}\right]^2 . \ \square$$

Note that if $\text{ARE}(S,T) > 1$, then S is better than T, while $\text{ARE}(S,T) < 1$ implies T is superior to S. This comparison depends only on $F(\cdot)$ and not on n, α or θ_1. So a clear comparison is possible.

To illustrate the use of ARE's, we compare the Wilcoxon signed rank test (denoted S^+) as described in (3.4) to the one-sample t-test (denoted T^+) as described in (1.1). Some values of their ARE are displayed in Table 6.3. We see that while the Wilcoxon signed rank test is not quite as good as the t-test when the underlying population is normal, in the other cases it is as good or better. If the underlying distribution is heavy-tailed then the improved power can be considerable.

Table 6.3

Distribution	$\text{ARE}(S^+,T^+)$
Uniform	1.000
Normal	.955
Logistic	1.097
Double Exponential	1.500
Cauchy	$+\infty$

7. USE OF SCORES. A rank test can be selected to have excellent power properties if the nature of the underlying distribution is known. Different rank tests are constructed by using a score function $\varphi(\cdot)$ in the description of the rank test. Again we illustrate the possibilities in the one-sample location problem. Here, a signed rank statistic with a score function $\varphi(\cdot)$ is of the form

$$S = \sum_{i=1}^{n} \Psi_i \varphi\left(\frac{R_i^+}{n+1}\right). \tag{7.1}$$

Different choices for $\varphi(\cdot)$, defined over the interval $(0,1)$, alter the nature of the resulting tests. For example, if $\varphi(u) = u$, the test statistic S is just $(n+1)^{-1}S^+$, and the test is equivalent to the Wilcoxon signed rank test. If $\varphi(u) \equiv 1$, then S becomes

$$B = \sum_{i=1}^{n} \Psi_i ,$$

a statistic which merely counts the number of positive X_i observations. The test based on this intuitive statistic is called the <u>sign test</u> and it is known to have good power properties when the underlying distribution is quite heavy-tailed. Two other possible score functions are illustrated in Table 7.2 and the ARE's of the resulting tests, relative to the Wilcoxon signed rank test are displayed in Table 7.3.

Table 7.2

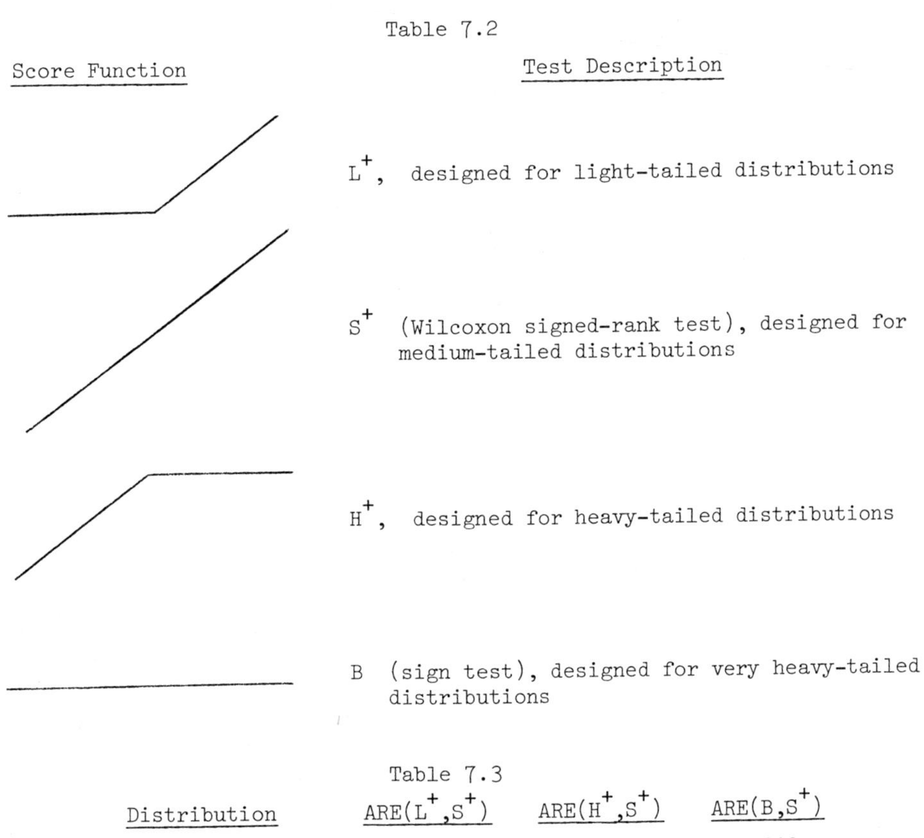

Score Function	Test Description
	L^+, designed for light-tailed distributions
	S^+ (Wilcoxon signed-rank test), designed for medium-tailed distributions
	H^+, designed for heavy-tailed distributions
	B (sign test), designed for very heavy-tailed distributions

Table 7.3

Distribution	$ARE(L^+,S^+)$	$ARE(H^+,S^+)$	$ARE(B,S^+)$
Uniform	2.000	.500	.333
Normal	.927	.870	.667
Logistic	.781	.945	.750
Double Exponential	.500	1.125	1.333
Cauchy	.264	1.339	1.333

8. BOOKS FOR FURTHER READING.

Methods:

Hollander, M. and Wolfe, D.A. (1973). Nonparametric Statistical Methods. Wiley, New York.

Methods plus some theory:

Lehmann, E.L. (1975). Nonparametrics: Statistical Methods Based on Ranks. Holden-Day, San Francisco.

Theory plus some methods:

Randles, R.H. and Wolfe, D.A. (1979). Introduction to the Theory of Nonparametric Statistics. Wiley, New York.

Theory:

Hájek, J. and Šidák, Z. (1967). Theory of Rank Tests. Academic Press, New York.

DEPARTMENT OF STATISTICS
THE UNIVERSITY OF IOWA
IOWA CITY, IOWA 52242

Proceedings of Symposia in Applied Mathematics
Volume 23
1980

RANK ESTIMATES FROM NONPARAMETRIC TESTS

Thomas P. Hettmansperger

ABSTRACT. For testing hypotheses about a location parameter such as the population mean or median or about differences in location parameters, nonparametric tests can be considered as alternatives to the classical t and F tests. Unlike the classical tests, nonparametric tests have significance levels which are independent of the distribution of the underlying population. Since the systematic development and study of nonparametric tests began around 1945, much has been learned about their excellent efficiency and power properties. There is strong evidence to suggest that unless the underlying population distribution is known very precisely, nonparametric tests are to be preferred over the classical t and F tests.

In 1963, Hodges and Lehmann derived estimates of location from rank tests. These so-called R-estimates inherit the efficiency properties of the parent rank tests. This paper will concentrate on the development of R-estimates and their properties.

Beginning in the middle 1960's, there has been much interest in the area of robust estimation. A robust estimate is not highly sensitive to a small fraction of the data and in particular, the sample mean is not robust. We will also develop R-estimates from the principles of robust estimation. An extension to rank estimates in regression will be given.

1. THE WILCOXON SIGNED RANK TEST. We begin with a nonparametric test of hypothesis in the one sample location model. Suppose $X_1, \ldots, X_n$ is a random sample (independent, identically distributed random variables) from a continuous, symmetric distribution. To represent this model we select a distribution function $F(x)$ from

$$\Omega_s = \{F: \ F(x) \text{ continuous and } F(x) + F(-x) = 1\}.$$

The class Ω_s contains all continuous, symmetric distributions centered at 0. Then the distribution sampled is $F(x - \theta)$ so that θ represents the point of symmetry or location of the sampled distribution.

We wish to test H_0: $\theta = 0$ vs. H_A: $\theta \neq 0$ with the Wilcoxon Signed Rank test which is given by

1980 Mathematics Subject Classification 62G05

$$S = \sum_{i=1}^{n} a(R_i^+) \operatorname{sgn} X_i$$

where R_i^+ is the rank of $|X_i|$ among $|X_1|, \ldots, |X_n|$ and the rank score $a(i)$ is determined by a non-decreasing score generating function $\phi(u)$, $0 < u < 1$ in the following manner:

$$a(i) = \phi(\frac{i}{n+1}).$$

There are many possibilities for $\phi(u)$; however, we will restrict attention to $\phi(u) = 3^{1/2}u$. This yields the Wilcoxon Signed Rank statistic in a form which is convenient for studying its properties. Note that S is a standardized form of the excess of the sum of ranks of the positive observations over the sum of ranks of the negative observations when absolute values are ranked.

Note 1.1 Under H_0: $\theta = 0$ it follows that

(i) $ES = 0$

(ii) $\operatorname{Var} S = \frac{n(2n+1)}{2(n+1)} \sim n$

(iii) $n^{-1}S \to 0$ in probability

(iv) $n^{-1/2}S \to Z$, in distribution, where Z is a standard normal random variable.

Note 1.2 The size α test of hypothesis is determined by:

Reject H_0: $\theta = 0$ if $|S| > c$ where $P_{H_0}(|S| > c) = \alpha$.

From (iv) the critical value c can be approximated, using the normal distribution, by $c \doteq n^{1/2} z_{\alpha/2}$ where

$$\alpha = P_{H_0}(|S| > c) = P_{H_0}(n^{-1/2}|S| > n^{-1/2}c) \doteq P(|Z| > z_{\alpha/2})$$

and $z_{\alpha/2}$, found in the standard normal table, is the upper critical point for a two sided test. If $\Phi(\cdot)$ is the standard normal cdf then $1 - \Phi(z_{\alpha/2}) = \alpha/2$.

2. THE COUNTING FORM. Let $\psi_i = 1$ or 0 as $X_i > 0$ or otherwise and define

$$T = \sum_{i=1}^{n} R_i^+ \psi_i,$$

the sum of ranks of the positive observations among the absolute values. Then

$$T = \# \frac{X_i + X_j}{2} > 0 \qquad i \le j.$$

The set of $n(n+1)/2$ pairwise averages $(X_i + X_j)/2$, $i \le j$, is called the set of Walsh averages. Hence T is the number of positive Walsh averages.

Note 2.1

$$S = \frac{3^{1/2}}{n+1}\left[\#\frac{X_i + X_j}{2} > 0 - \#\frac{X_i + X_j}{2} < 0\right]$$

$$= \frac{2(3^{1/2})}{n+1} T - \frac{n(3^{1/2})}{2} .$$

3. THE GRAPHICAL REPRESENTATION. Define the function $S(\theta)$ by

$$S(\theta) = \Sigma\, a(R_i^+(\theta))\ \mathrm{sgn}(X_i - \theta)$$

where $R_i^+(\theta)$ is the rank of $|X_i - \theta|$ among $|X_1 - \theta|, \ldots, |X_n - \theta|$. Hence from Note 2.1 we have

$$\frac{n+1}{3^{1/2}} S(\theta) = \left[\#\frac{X_i + X_j}{2} > \theta - \#\frac{X_i + X_j}{2} < \theta\right]$$

$$= \left[2\left(\#\frac{X_i + X_j}{2} > \theta\right) - \frac{n(n+1)}{2}\right] .$$

Note that $S(0) = S$, the Wilcoxon Signed Rank statistic. From this representation it follows that the graph of $S(\theta)$ is a non-increasing step function with steps at the $n(n+1)/2$ Walsh averages. See the figure.

4. THE POINT AND INTERVAL ESTIMATES. We can now derive an estimate of θ from the test of hypothesis. As in Note 1.1, when θ is the true value of the parameter it follows that

$$E_\theta S(\theta) = 0.$$

Hence we take as our estimate $\hat{\theta}$ such that

$$S(\hat{\theta}) \doteq 0.$$

We have approximately 0 because of the discreteness of the graph. The graph either jumps across 0 at the middle Walsh average or it is 0 for all θ values between the middle two Walsh averages. Hence in either case

$$\hat{\theta} = \underset{i \leq j}{\mathrm{med}} \frac{X_i + X_j}{2} .$$

This was first suggested by Hodges and Lehmann (1963).

Further we can determine an approximate $(1-\alpha)100\%$ confidence interval for θ by matching $n^{-1/2}\ S(\theta)$ to $Z_{\alpha/2}$ and $-Z_{\alpha/2}$. Hence the confidence interval $[\hat{\theta}_L, \hat{\theta}_U]$ is determined by

$$S(\hat{\theta}_L) = n^{1/2}\ Z_{\alpha/2} \text{ and } S(\hat{\theta}_U) = -n^{1/2}\ Z_{\alpha/2}.$$

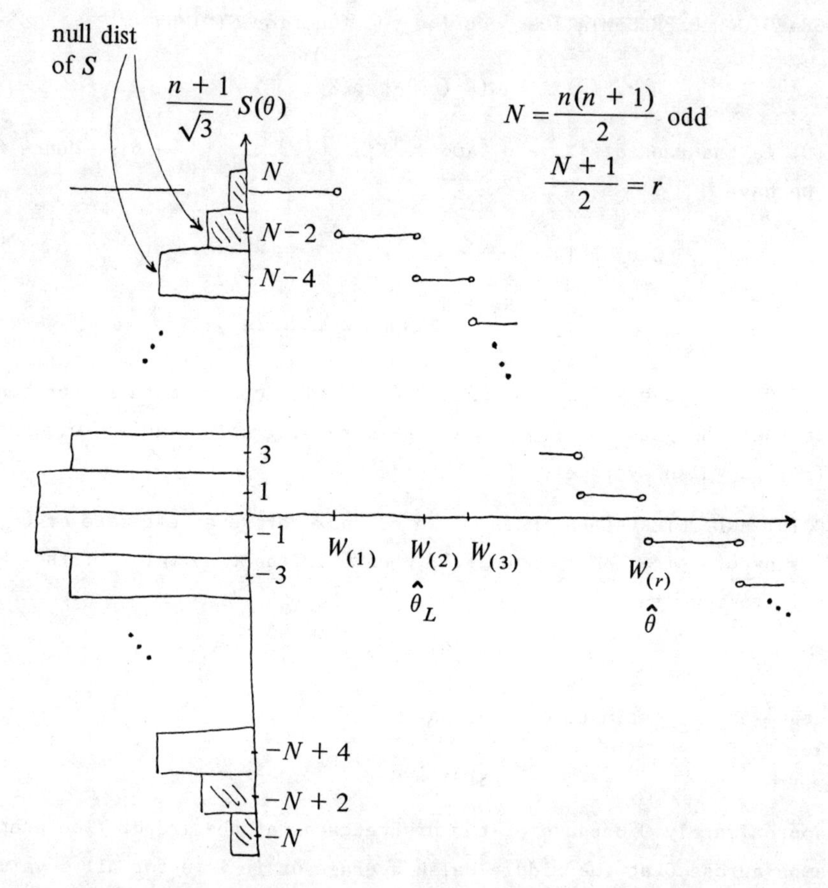

$$1 - \alpha = P_\theta\left(-N + 2k \leqslant \frac{n+1}{\sqrt{3}} S(\theta) \leqslant N - 2k\right) = P_\theta\left(W_{(k)} < \theta < W_{(N-k+1)}\right)$$

The example is for $k = 2$; critical region is shaded.

FIGURE

From the graph, $\hat{\theta}_L$ and $\hat{\theta}_U$ can be identified with a specific pair of ordered Walsh averages, where $W_{(1)} \leq \ldots \leq W_{(N)}$ are the ordered Walsh averages:

$$\hat{\theta}_L = W_{(k)} \quad \text{and} \quad \hat{\theta}_U = W_{(N-k+1)}$$

where $N = n(n+1)/2$ and $k \doteq N/2 - Z_{\alpha/2}(n+1)n^{1/2}/2(3)^{1/2}$ from Note 1.1.

At this point, given a set of data, we have available three basic forms of statistical inference: an hypothesis test, a point estimate and a confidence interval. The three are interrelated with the point and interval estimates being derived from the hypothesis test. The interrelationship is illustrated in the figure. Not surprisingly they share efficiency properties. Hence if it is established that, according to some criterion, the test is a good one then the point and interval estimates are also good.

We will study the power and efficiency of the test and estimates for large n; that is, asymptotically as n tends to infinity. One of the most powerful techniques that mathematics has provided the applied sciences is the linear approximation. In the next section we will develop a linear approximation of $n^{-1}S(\theta)$. This approximation will enable us to find the asymptotic power and efficiency of the test, the asymptotic distribution and efficiency of the point estimate and the asymptotic length of the confidence interval. From these results we will make a comparison of these rank procedures with the classical t-test, estimate and confidence interval.

5. THE SLOPE. For large n, $n^{-1} S(\theta)$ is approximately constant for each θ with high probability. In this case the step sizes shrink and at the center of the graph the Walsh averages become quite dense. Without loss of generality we take $\theta = 0$ as the true parameter value. We will approximate the step function with a straight line. Since the standard deviation of $n^{-1} S(\theta)$ behaves like $n^{-1/2}$ (see Note 1.1) we can study the stochastic behavior of $n^{-1} S(\theta)$ on intervals like $[n^{-1/2}a, n^{-1/2}b]$, $a < b$.

$$\text{Slope} = \frac{n^{-1}S(n^{-1/2}b) - n^{-1}S(n^{-1/2}a)}{n^{-1/2}b - n^{-1/2}a}$$

$$= \frac{-\,2(3)^{1/2}[n(n+1)]^{-1}\,\Sigma\,\Sigma\,I_{ij}}{n^{-1/2}b - n^{-1/2}a}$$

where $I_{ij} = 1$ if $(x_i + x_j)/2 \;\varepsilon [n^{-1/2}a, n^{-1/2}b]$ and 0 otherwise.

Hence

$$E(\text{Slope}) \sim -\,3^{1/2}\,\frac{[G(n^{-1/2}b) - G(n^{-1/2}a)]}{n^{-1/2}b - n^{-1/2}a}$$

and

$$E(\text{Slope}) \to -\,3^{1/2}\,g(0)$$

where $G(t) = P(\frac{X_1 + X_2}{2} \leq t)$ and $g(t) = G'(t)$, the density of $G(t)$.

Under the assumption that the true value of θ is 0 the distribution G is related to the sampled, symmetric distribution F by

$$G(t) = \int F(2t - x)\, f(x)\, dx$$

$$g(t) = 2 \int f(2t - x)\, f(x)\, dx$$

$$g(0) = 2 \int f^2(x)\, dx.$$

Hence

$$E(\text{Slope}) \to -(12)^{1/2} \int f^2(x)\, dx.$$

<u>Note 5.1</u> The approximation can be written as follows

$$\frac{n^{-1}S(n^{-1/2}b) - n^{-1}S(n^{-1/2}a)}{n^{-1/2}b - n^{-1/2}a} \doteq -(12)^{1/2} \int f^2(x)\, dx$$

or more precisely

$$n^{-1/2}S(n^{-1/2}b) - n^{-1/2}S(n^{-1/2}a) = -(b-a)(12)^{1/2} \int f^2(x)\, dx + o_p(1).$$

The term $o_p(1)$ tends to 0 in probability uniformly for all a, b such that $|b-a| < K$ for any constant K. The details are worked out by Lehmann (1963) and involve only an application of Chebyshev's inequality with uniform bounds on variances and covariances.

<u>Note 5.2</u> If we let $a = n^{1/2}\theta_0$ and $b = n^{1/2}\theta$ then the approximation becomes

$$n^{-1/2}S(\theta) = n^{-1/2}S(\theta_0) - (\theta - \theta_0)\, n^{1/2}(12)^{1/2} \int f^2(x)\, dx + o_p(1)$$

which expresses the linear approximation as an expansion of $n^{-1/2}S(\theta)$ about the point θ_0.

6. THE ASYMPTOTIC POWER. For testing H_0: $\theta = 0$ against H_A: $\theta > 0$ we reject H_0: $\theta = 0$ if $n^{-1/2}S(0) > Z_\alpha$ where $P_0(n^{-1/2}S(0) > Z_\alpha) \doteq \alpha = 1 - \Phi(Z_\alpha)$ determines the approximate size α critical value Z_α. (Recall $\Phi(\cdot)$ denotes the standard normal distribution function.)

For a fixed alternative θ the power $P_\theta(n^{-1/2}S(0) > Z_\alpha)$ will tend to 1; hence, we have a consistent test. In order to stabilize the asymptotic power as n increases we consider the power along a sequence of alternative $\{\theta_n\}$, $\theta_n = n^{-1/2}\theta$.

From the linear approximation (Note 5.1) with $a = -\theta$ and $b = 0$ we have

$$\begin{aligned} P_{\theta_n}(n^{-1/2}S(0) > Z_\alpha) &= P_0(n^{-1/2}S(-n^{-1/2}\theta) > Z_\alpha) \\ &\sim P_0(n^{-1/2}S(0) + \theta(12^{1/2}) \int f^2 > Z_\alpha) \\ &= P_0(n^{-1/2}S(0) > Z_\alpha - \theta(12^{1/2}) \int f^2). \end{aligned}$$

Hence the approximate or asymptotic power of the Wilcoxon Signed Rank test along the sequence of alternatives $\{\theta_n\}$ is

$$P_{\theta_n}(n^{-1/2}S(0) > Z_\alpha) \doteq 1 - \Phi(Z_\alpha - \theta(12^{1/2}) \int f^2).$$

This asymptotic power is large for distributions F such that the slope $(12)^{1/2} \int f^2$ is large.

For the t-test, $t = n^{1/2}\overline{X}/s$, we let $t(\theta) = n^{1/2}(\overline{X} - \theta)/s$ where s is the sample standard deviation. Note that $t(\theta)$ is linear in θ with slope $-1/s$ where s converges to σ_F in probability. It can then be shown that

$$P_{\theta_n}(t > Z_\alpha) \doteq 1 - \Phi(Z_\alpha - \theta/\sigma_F).$$

Hence to compare the Wilcoxon Signed Rank test to the t-test we will compare $(12)^{1/2} \int f^2$ to $1/\sigma_F$. In the next section we will show that the quantity of interest is the square of their ratio: $\sigma_F^2\, 12 \,(\int f^2)^2$.

7. PITMAN EFFICIENCY. Let $C_S = (12)^{1/2} \int f^2$ and $C_t = \sigma_F^{-1}$. We choose θ_1 and θ_2 such that $\theta_1 C_S = \theta_2 C_t$ so that for asymptotically size α tests the asymptotic powers are the same. In addition we match the sequences of alternatives so that $n^{-1/2}\theta_1 \sim m^{-1/2}\theta_2$. (See Lehmann (1975) for details.) Then

$$\frac{m}{n} \sim \left(\frac{\theta_2}{\theta_1}\right)^2 = \left(\frac{C_S}{C_t}\right)^2 .$$

The limiting ratio of sample sizes required to have the same asymptotic level and power along the same sequence of alternatives is called the Pitman efficiency. The efficiency of S relative to t is then given by

$$e(S, t) = \left(\frac{C_S}{C_t}\right)^2 = 12\, \sigma_F^2 \,(\int f^2)^2 .$$

The quantities C_S^2 and C_t^2 are called the efficacies of the two tests and are simply the squares of the slopes.

8. THE ASYMPTOTIC DISTRIBUTION AND EFFICIENCY OF $\hat{\theta}$. The Hodges-Lehmann estimate $\hat{\theta}$ was derived from S in Section 4. We now use the asymptotic linearity to derive its asymptotic distribution. From Section 4 and Note 5.2 if we let θ_0 denote the true parameter value then we can write:

$$n^{1/2}S(\hat{\theta}) \doteq 0$$

$$n^{-1/2}S(\theta_0) - (12)^{1/2} \int f^2 \, n^{1/2}(\hat{\theta} - \theta_0) \doteq 0$$

$$n^{1/2}(\hat{\theta} - \theta_0) \doteq \frac{1}{(12)^{1/2} \int f^2} n^{-1/2}S(\theta_0).$$

From Note 1.1, $n^{-1/2}S(\theta_0)$ converges in distribution to the standard normal and hence

$$n^{1/2}(\hat{\theta} - \theta_0) \to V \text{ in distribution}$$

where V is distributed as a normal with mean 0 and variance $1/12(\int f^2)^2$. Hence the asymptotic variance of $n^{1/2}\hat{\theta}$ is the reciprocal square of the slope.

Since the asymptotic variance of $n^{1/2}\overline{X}$ is σ_F^2, we immediately have that the asymptotic efficiency of $\hat{\theta}$ relative to $\overline{X}$ is

$$e(\hat{\theta}, \overline{X}) = \frac{\text{asy var } \overline{X}}{\text{asy var } \hat{\theta}}$$

$$= 12\, \sigma_F^2 (\int f^2)^2 = e(S, t).$$

In general the asymptotic efficiency of two estimates, given as the reciprocal ratio of the asymptotic variances, is equal to the Pitman efficiency of the tests from which they are derived.

9. ASYMPTOTIC LENGTH OF THE CONFIDENCE INTERVAL. Again from Section 4 the interval is defined by $n^{-1/2}S(\hat{\theta}_L) = Z_{\alpha/2}$ and $n^{-1/2}S(\hat{\theta}_U) = -Z_{\alpha/2}$ where $1 - \Phi(Z_{\alpha/2}) = \alpha/2$. From Note 5.1 we have

$$n^{-1/2}S(\hat{\theta}_U) - n^{-1/2}S(\hat{\theta}_L) + n^{1/2}(\hat{\theta}_U - \hat{\theta}_L)(12)^{1/2} \int f^2 \doteq 0.$$

Hence

$$\frac{n^{1/2}(\hat{\theta}_U - \hat{\theta}_L)}{2\, Z_{\alpha/2}} \to \frac{1}{(12)^{1/2} \int f^2} \text{ in probability}$$

and the standardized length of the confidence interval converges to the reciprocal of the slope, in probability.

In a similar fashion for the length of the t-interval we have, letting L(t) denote the length,

$$\frac{n^{1/2} L(t)}{2\, Z_{\alpha/2}} \to \sigma_F \text{ in probability.}$$

Hence the squared ratio of slopes is a measure of the relative efficiency of the two intervals and is again given by e(S, t).

10. REMARKS ON EFFICIENCY. Statements about $12\,\sigma_F^2(\int f^2)^2$ directly reflect the comparison of the Wilcoxon Signed Rank test, Hodges-Lehmann estimate and confidence interval with the t-test, $\overline{X}$ and t-interval.

If F is the normal distribution then $12\,\sigma_F^2(\int f^2)^2 = .955$. Hence if the underlying distribution is normal so that we should be using a t-procedure we actually pay very little for erroneously using a rank procedure. If F is the double exponential or Laplace distribution, which has heavier tails than the normal, $12\,\sigma_F^2(\int f^2)^2 = 1.5$. This reflects a general tendency of rank procedures to be more efficient than t-procedures for heavy tailed distributions. In 1956 Hodges and Lehmann showed that the infimum of $12\,\sigma_F^2(\int f^2)^2$ over all symmetric distributions is .864. Hence the rank procedures are never very inefficient and may be much more efficient than the t-procedures.

11. WEIGHTED LEAST ABSOLUTE DEVIATIONS. For estimating the location parameter θ least squares entails minimizing $\Sigma(X_i - \theta)^2$ or solving $\Sigma(X_i - \theta) = 0$ which yields $\overline{X}$ the sample mean. Least squares can be criticized on the grounds that extreme observations exert too much influence on the estimate. This influence enters through the quadratic tail of the function that is minimized. To alleviate this sensitivity to extreme values we could minimize $\Sigma|X_i - \theta|$ or solve $\Sigma\ \mathrm{sgn}(X_n - \theta) \doteq 0$ to get the sample median. Some would say that replacing the square with the absolute value is an overreaction especially since the median is not very efficient relative to the mean at the normal distribution. (The efficiency is .64.) Hence we consider weighted least absolute deviation estimates which hopefully (a) are fairly efficient relative to the mean at the normal distribution and (b) are not nearly as sensitive to extreme observations as the mean. In other words we give up some efficiency at the normal to gain stability near the normal.

We define our estimate by minimizing

$$D(\theta) = \Sigma\ a(R_i^+(\theta))|X_i - \theta|$$

where $R_i^+(\theta)$ is the rank of $|X_i - \theta|$ among the absolute values. Hence the absolute deviation $|X_i - \theta|$ is weighted by a function of its relative size expressed through its rank. The function $D(\theta)$ is a measure of dispersion and the estimate minimizes this dispersion. We can differentiate $D(\theta)$, except at a finite set of points, to get

$$-\ S(\theta) = \Sigma\ a(R^+(\theta))\mathrm{sgn}(X_i - \theta) \doteq 0.$$

Hence if $a(i) = 3^{1/2} i/(n+1)$ given in Section 1, the weighted least absolute deviation estimate is just the Hodges and Lehmann estimate $\hat{\theta}$ given in Section 4. Note that $\hat{\theta}$ is quite efficient relative to $\overline{X}$ at the normal. (The efficiency is .955 from Section 10.) Thus the first criteria (a) is met and we turn next to the sensitivity of $\hat{\theta}$ to extreme observations.

12. THE INFLUENCE CURVE. Recall from Section 1 that we have $X_1, \ldots, X_n$, a random sample from $H(x) = F(x - \theta)$ for some F in Ω_s. The mean functional $T(H) = \int x dH(x)$ is equal to the point of symmetry of the underlying distribution. The estimate is then $T(H_n) = \int x dH_n(x) = \overline{X}$ where $H_n(x)$ is the empirical distribution function. ($H_n(x)$ assigns mass n^{-1} to each observation.) The sensitivity of the sample mean $\overline{X}$ to extreme observations is expressed through the instability of the mean functional $T(H) = \int x dH(x)$.

Hampel (1974) introduced the Gateaux derivative of a general functional $T(H)$ in the direction of a point mass distribution. Formally we let $\delta_y(x) = 0$ if $x < y$ and 1 if $x \geq y$ and compute

$$I(y) = \frac{d}{d\varepsilon} T[(1-\varepsilon)H + \varepsilon\delta_y]\Big|_{\varepsilon=0} .$$

This is called the influence curve and measures the rate of change of the functional $T(\cdot)$ at H in the direction of δ_y. This measures the effect of "tossing in" an extreme observation y.

Since the actual value of the location parameter θ has no effect on the influence, without loss of generality we will let $\theta = 0$ and replace H by F. Let $F_\varepsilon = (1-\varepsilon)F + \varepsilon\delta_y$ then for the mean functional

$$T(F_\varepsilon) = \int x dF_\varepsilon(x) = \varepsilon \int x d\delta_y(x) = \varepsilon y.$$

Hence $I(y) = y$ and the instability of the mean is expressed by noting it has <u>unbounded</u> influence.

The Hodges-Lehmann estimate $\hat{\theta}$ is the median of the Walsh averages. The relevant functional is the median of the distribution of $(X_1 + X_2)/2$. From Section 5 we have the distribution function given by $G(t) = \int H(2t-x)dH(x)$ and the median functional $T(H)$ is defined implicitly by

$$\int H(2T(H) - x)dH(x) = 1/2.$$

Again let $\theta = 0$ and replace H by F. Now insert F_ε into the above definition of the median and differentiate with respect to ε. At $\varepsilon = 0$ we find

$$I(y) = \frac{F(y) - 1/2}{\int f^2(t)dt} .$$

Note that the distribution function F is bounded between 0 and 1; hence, unlike the mean the Hodges-Lehmann estimate has a <u>bounded</u> influence curve. For this reason extreme observations cannot have an arbitrary effect on the estimator. Hence the second criterion (b) in Section 11 is satisfied by $\hat{\theta}$.

Based on the influence curves and other measures of robustness discussed by Hampel (1974) we conclude that the Hodges-Lehmann estimate $\hat{\theta}$ is preferable to the classical estimate $\overline{X}$.

13. RANK ESTIMATION IN THE REGRESSION MODEL. In the location model we suppose that the observation X_i can be written as $X_i = \theta + e_i$ where $e_1, \ldots, e_n$ is a random sample from $F \in \Omega_s$. Hence $Ee_i = 0$ and $EX_i = \theta$ when expectations exist. In the regression model we suppose that the mean EX_i is more complicated. The most natural extension from the constant θ is to replace θ by a linear function for each i:

$$EX_i = a + bc_i, \; i = 1, \ldots, n.$$

We assume $c_1, \ldots, c_n$ are known constants (perhaps set by the experimenter) and the unknown parameters are a and b. This simple regression model, containing only one independent variable, c_i, has many applications. For example c_i might be the SAT score and X_i the GPA at the end of the first semester. The problem is to estimate the mean GPA for different values of the SAT score.

The least squares estimates of a and b are determined by minimizing $\Sigma(X_i - a - bc_i)^2$. They have the same instability problems as the mean discussed in Sections 11 and 12. We next introduce a measure of dispersion similar to $D(\theta)$ discussed in Section 11.

We will rank the residuals $X_1 - a - bc_1, \ldots, X_n - a - bc_n$. Note that the ranks are invariant to the intercept a so we actually only need to rank $X_1 - bc_1$, $\ldots, X_n - bc_n$. Let $R_1(b), \ldots, R_n(b)$ denote these ranks and define

$$V(b) = \Sigma[R_i(b) - \tfrac{n+1}{2}](X_i - bc_i).$$

Hence we consider a linear combination of the residuals where the weights are the centered ranks. Just as in the one sample case, Section 1, other functions of the ranks could be considered; however, the centered ranks generate the Wilcoxon procedures. Further we can differentiate V(b) except at a finite set of points to get

$$- U(b) = \Sigma\, c_i[R_i(b) - \tfrac{n+1}{2}].$$

The rank estimate of b, say $\hat{b}$, minimizes V(b) and solves $U(b) \doteq 0$. Generally numerical methods have to be used to compute $\hat{b}$. See Jaeckel (1972).

To estimate the intercept a we first form the estimated residuals $X_1 - \hat{b}c_1$, $\ldots, X_n - \hat{b}c_n$ and then apply the one sample methods. Hence $\hat{a}$ is the median of the Walsh averages of the residuals.

In the case of least squares the estimates $\hat{a}_{LS}$ and $\hat{b}_{LS}$ have variances equal to $\sigma_F^2\, \Sigma\, c_i^{\,2}/n\Sigma(c_i - \bar{c})^2$ and $\sigma_F^2/\Sigma(c_i - \bar{c})^2$, respectively. For the rank estimates $\hat{a}$ and $\hat{b}$ we have the same formulas asymptotically with σ_F^2 replaced by $12(\int f^2)^2$. These results along with the asymptotic normality of the estimates can be derived from a linear approximation of U(b) similar to that of Section 5.

14. THE TWO SAMPLE LOCATION PROBLEM. In Section 13 let $c_i = 1$, for $i = 1, \ldots,$ m and 0 for $i = m+1, \ldots, n$. Then $X_1, \ldots, X_m$ represents the first sample and $X_{m+1}, \ldots, X_n$ represents the second sample. The parameter b represents the difference in their locations since

$$EX_i = \begin{cases} a+b & i = 1, \ldots, m \\ a & i = m+1, \ldots, n. \end{cases}$$

In this case $-U(b) = \sum_1^m [R_i(b) - (n+1)/2]$ is the centered sum of ranks of the first sample when the combined sample is ranked. The test statistic $-U(0)$ is the Mann-Whitney-Wilcoxon Rank statistic.

Note that

$$\begin{aligned} -U(0) &= \sum_1^m R_i(0) - \frac{m(n+1)}{2} \\ &= \#(X_i - X_j > 0) - \frac{m(n-m)}{2} \qquad \begin{matrix} i = 1, \ldots, m \\ j = m+1, \ldots, n \end{matrix} \end{aligned}$$

and

$$-U(b) = \#(X_i - X_j > b) - \frac{m(n-m)}{2} \qquad \begin{matrix} i = 1, \ldots, m \\ j = m+1, \ldots, n. \end{matrix}$$

This follows from the fact that the rank $R_i(0)$ is equal to the number of times items in the first sample are less than or equal to X_i plus the number of times items in the second sample are less than X_i.

Now along the lines of Section 4 we can show that the estimate $\hat{b}$ of the difference in locations is determined by

$$-U(b) \doteq 0$$

and hence

$$\hat{b} = \operatorname{med}(X_i - X_j) \qquad \begin{matrix} i = 1, \ldots, m \\ j = m+1, \ldots, n, \end{matrix}$$

the median of the m(n-m) differences across the two samples. This was proposed by Hodges and Lehmann (1963).

BIBLIOGRAPHY

1. Hampel, F. R., "The influence curve and its role in robust estimation", J. Amer. Statist. Assoc., 69, (1974), 383-393.

2. Hettmansperger, T. P. and Utts, J. M., "Robustness properties for a simple class of rank estimates", Commun. Statist.-Theor. Meth., A6(9), (1977), 855-868.

3. Hodges, J. L., Jr. and Lehmann, E. L., "The efficiency of some non-parametric competitors of the t-test", Ann. Math. Statist., 27, (1956), 324-335.

4. Hodges, J. L., Jr. and Lehmann, E. L., "Estimates of location based on rank tests", Ann. Math. Statist., 34, (1963), 598-611.

5. Huber, P. J. Robust Statistical Procedures, SIAM, Philadelphia, PA, 1977.

6. Jaeckel, L. A., "Estimating regression coefficients by minimizing the dispersion of the residuals", Ann. Math. Statist., 43 (1972), 1449-1458.

7. Lehmann, E. L., "Nonparametric confidence intervals for a shift parameter", Ann. Math. Statist., 34 (1963), 1507-1512.

8. Lehmann, E. L., Nonparametrics: Statistical Methods Based on Ranks, Holden-Day Inc., San Francisco, CA, 1975.

9. Randles, R. H. and Wolfe, D. A., Introduction to the Theory of Non-parametric Statistics, Wiley-Interscience, New York, New York, 1979.

Department of Statistics
Pond Lab
The Pennsylvania State University
University Park, PA 16802

Proceedings of Symposia in Applied Mathematics
Volume 23
1980

STATISTICAL INFERENCES FOR ORDERED PARAMETERS: A PERSONAL VIEW OF ISOTONIC REGRESSION SINCE THE WORK BY BARLOW, BARTHOLOMEW, BREMNER AND BRUNK

Tim Robertson and F. T. Wright

1. A PROTOTYPAL PROGRAM. The basic concepts in an order restricted inference problem are presented in the following bioassay example. The parameters of interest are p_i, the probability of a particular response, say R, to a drug administered at level s_i, for $i = 1,2,\ldots,k$. To obtain estimates of the parameters, the drug is administered at level s_i to n_i animals and the experimental results, x_{ij}, $j = 1,2,\ldots,n_i$, are recorded for each i. If the jth animal given the drug at level s_i exhibits R, then $x_{ij} = 1$ and $x_{ij} = 0$ otherwise. The likelihood function is

$$L(p_1,p_2,\ldots,p_k) = \prod_{i=1}^{k} p_i^{n_i\hat{p}_i}(1-p_i)^{n_i(1-\hat{p}_i)}$$

where $\hat{p}_i = \sum_{j=1}^{n_i} x_{ij}/n_i$ is the relative frequency estimate of p_i. So

$$(1)\quad -\ln L(p_1,p_2,\ldots,p_k) = \sum_{i=1}^{k} -\{n_i\hat{p}_i \ln p_i + n_i(1-\hat{p}_i)\ln(1-p_i)\}$$

and the maximum likelihood estimates (m.l.e.s) of $p_1,p_2,\ldots,p_k$ are obtained by minimizing (1).

If we have no further information concerning these parameters, then the minimum is obtained by minimizing each term separately, that is the minimum occurs at $p_i = \hat{p}_i$ for $i = 1,2,\ldots,k$. (The problem of minimizing each given term is just that of finding the m.l.e. of a proportion. But, of course, this is the sample proportion.) However, if it is believed that

$$(2)\quad p_i = f(s_i) \qquad i = 1,2,\ldots,k,$$

where the functional form of f is known but some of the constants determining f are unknown, then the likelihood function could be expressed in terms of these unknown constants and their m.l.e.s could be obtained. For instance, if for the dosage levels considered, one could assume that $p_i = [1+\exp\{-bs_i-a\}]^{-1}$, $i = 1,2,\ldots,k$, then the m.l.e.s of the p_i could be obtained by finding the m.l.e.s of a and b. If the number of constants needed to determine f is small, then these estimates are more accurate than the "raw" estimates $\hat{p}_i$ provided assumption (2) is correct or at least nearly so. If assumption (2) is incorrect

1980 Mathematics Subject Classification 62G05

the estimates based on this assumption may be unsatisfactory. So one might consider assumptions weaker than (2). Of course, it is expected that the estimates based on these weaker assumptions will not be as accurate as those based on (2) if in fact (2) holds. On the other hand, they should provide some protection in case (2) does not hold.

If the levels s_i are nondecreasing, then it might be reasonable to assume that the p_i are nondecreasing also and if this is so we say that the p_i are isotone with respect to the usual order on $S = \{s_1, s_2, \ldots, s_k\}$. (Since the order is understood we will simply say that the p_i are isotone.) Ayer, et al. (1955) obtained the m.l.e.s for the p_i subject to the restriction $p_1 \le p_2 \le \ldots \le p_k$. The "pool adjacent violators algorithm" (PAVA) provides the easiest means for computing these estimates subject to the isotonic restriction. If $\hat{p}_1 \le \hat{p}_2 \le \ldots \le \hat{p}_k$ then the minimum of (1) occurs at a point that satisfies the restriction and so the solution to the restricted minimization problem, $(\bar{p}_1, \bar{p}_2, \ldots, \bar{p}_k)$, is given by $\bar{p}_i = \hat{p}_i$ for $i = 1,2,\ldots,k$. If a violation of the order restrictions occurs, say $\hat{p}_i > \hat{p}_{i+1}$, then it can be shown that $\bar{p}_i = \bar{p}_{i+1}$, and so the two estimates are pooled. For A a nonempty subset of S, let $\sum_A$ denote $\sum_{\{i:\, s_i \varepsilon A\}}$ and set

$$M(A) = \textstyle\sum_A n_i \hat{p}_i / \sum_A n_i .$$

(Note that M(A) is the proportion of responses R in the sample obtained by combining the samples at level s_i for all $s_i \varepsilon A$.) If there are violators a tree is constructed as follows. Beginning with any violation, one replaces both of the violators with $M(\{s_i, s_{i+1}\})$. This procedure is repeated until a properly ordered set of values is obtained. If at some stage, $M(\{s_\alpha, s_{\alpha+1}, \ldots, s_\beta\}) > M(\{s_{\beta+1}, \ldots, s_\gamma\})$, then both of these values on the tree are replaced with $M(\{s_\alpha, \ldots, s_\gamma\})$. The estimate $\bar{p}_i$ is the final value derived from $\hat{p}_i$ by this process. We illustrate the PAVA by an example.

Example. Suppose that there are five levels with $p_1 \le p_2 \le \ldots \le p_5$ and suppose each n_i is 10. The $\hat{p}_i$ are given below.

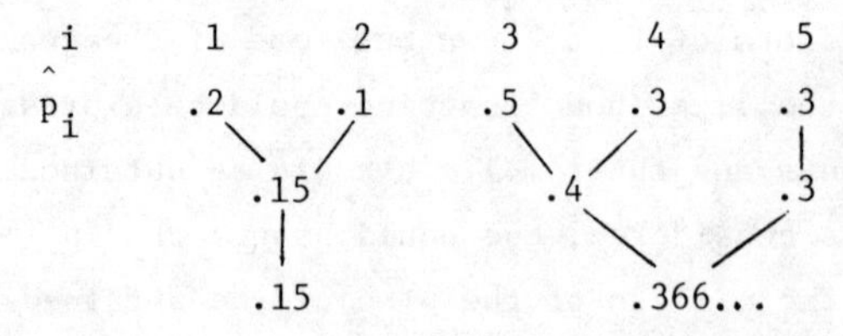

Note that initially there are two violations .2 > .1 and .5 > .3. After these are corrected by pooling there is still the violation .4 > .3. The estimates are $\bar{p}_1 = \bar{p}_2 = .15$ and $\bar{p}_3 = \bar{p}_4 = \bar{p}_5 = .366\ldots$.

Alternatively, these estimates can also be computed using the formula

(3) $\bar{p}_i = \max_{1 \le \alpha \le i} \min_{i \le \beta \le k} M(\{s_\alpha,\ldots,s_\beta\}),\ i = 1,2,\ldots,k.$

While the PAVA is easier to implement than this max-min formula, the latter has proved very useful in studying the properties of the estimators. There are several other algorithms for computing these estimates, some of which will be discussed later, but Barlow, et al. (1972) contains a thorough treatment of these estimators.

In some situations the researcher may not have the necessary information to impose a total order on the parameters of interest. In the bioassay problem we might be concerned not only with the dosage levels but also several methods for administering the drug, such as orally and by injections at various points in the body. We could use for the index set $S = \{(s_i(1),s_i(2)): i = 1,2,\ldots,k\}$ where the first coordinate indicates the level and the second coordinate the method of administration. Suppose that the levels and the methods of administration have been labelled numerically so that it would be reasonable to assume that $p_i \le p_j$ if $s_i(1) = s_j(1)$ and $s_i(2) \le s_j(2)$ or if $s_i(1) \le s_j(1)$ and $s_i(2) = s_j(2)$. In this case, S inherits its order relation from the natural partial order on R_2 and if $s_i << s_j$ (ie. $s_i(1) \le s_j(1)$ and $s_i(2) \le s_j(2)$) then $p_i \le p_j$. The p_i are said to be isotone with respect to the partial order $<<$ on S. Of course, one could envision examples in which $S \subset R_k$ and the order relation on S is inherited from the usual partial order on R_k. Another interesting partial order that might be imposed on the parameters is $p_1 \le p_2 \le \cdots \le p_{i_0} \ge p_{i_0+1} \ge \cdots \ge p_k$. Van Eeden (1956, 1957) studied the maximum likelihood estimation of probabilities subject to partial order restrictions.

When the order is not total, no notion of adjacency exists and the PAVA is not applicable. The algorithm most often used in its place is the minimum lower sets algorithm (MLSA) which is described below. If $<<$ is a partial order on a nonempty set S, then $L \subset S$ is a <u>lower layer</u> provided for $s,t \in S$ with $s << t$, $s \in L$ whenever $t \in L$. The collection of lower layers, $\mathcal{L}$, is closed under unions and intersections. (The complement of a lower layer is an <u>upper layer</u>.) We use the symbol L only to represent lower layers. The m.l.e.s of the p_i subject to the restriction that they are isotone with respect to $<<$ are given by

(4) $\bar{p}_i = \max_{\{L':\ s_i \notin L'\}} \min_{\{L:\ s_i \in L\}} M(L-L'),\ i = 1,2,\ldots,k.$

Notice that if the order is total, $s_1 \le s_2 \le \cdots \le s_k$, then the lower layers are of the form ϕ or $\{s_1,\ldots,s_\beta\}$ with $1 \le \beta \le k$ and (3) and (4) agree.

We now describe the MLSA. Set $L_0 = \phi$ and let L_1 be the largest lower layer (it can be shown that such exists) which minimizes $M(L-L_0)$ among all nonempty lower layers L. Next let L_2 be the largest lower layer which minimizes $M(L-L_1)$ among all lower layers L which properly contain L_1. Continuing this process we obtain $L_0 \subset L_1 \subset L_2 \subset \cdots \subset L_H = S$ for some positive integer H (recall S in finite). A given s_i is in $L_{\alpha+1} - L_\alpha$ for some α and $\bar{p}_i = M(L_{\alpha+1} - L_\alpha)$. The sets $L_{\alpha+1} - L_\alpha$ are called the level sets since the estimates are constant on them.

EXAMPLE. Suppose there are two levels and two methods of administering the drug, say orally and by injection, and $S = \{s_1 = (1,1), s_2 = (1,2), s_3 = (2,1), s_4 = (2,2)\}$. Let $n_1 = n_2 = n_3 = n_4 = 10$, $\hat{p}_1 = .3$, $\hat{p}_2 = .2$, $\hat{p}_3 = .1$ $\hat{p}_4 = .4$. The nonempty lower layers with their associated M values are $M(\{s_1\})= .3$, $M(\{s_1,s_2\}) = .25$, $M(\{s_1,s_3\}) = .2$, $M(\{s_1,s_2,s_3\}) = .2$ and $M(S) = .25$. So $L_0 = \phi$, $L_1 = \{s_1,s_2,s_3\}$ and then $L_2 = S$. Hence $\bar{p}_1 = \bar{p}_2 = \bar{p}_3 = .2$ and $\bar{p}_4 = .4$.

If the order on S is a total order, then the MLSA provides another means for computing the estimates in this case and it is not difficult to program.

2. ORDER RESTRICTED OPTIMIZATION. As we have seen the basic concepts in an order restricted inference problem are

an index set $S = \{s_1,s_2,\ldots,s_k\}$ with a partial order $<<$ on S,

a collection of parameters to be estimated, $\theta(s_i)$ for $i = 1,2,\ldots,k$, which are isotone with respect to $<<$,

independent random samples corresponding to the elements s_i, x_{ij} for $j = 1,2,\ldots,n_i$ and $i = 1,2,\ldots,k$, and

an objective function $D(\Phi)$ with domain the collection of real valued function on S (D also depends on the data x_{ij}).

One considers estimates $\bar{\theta} = (\bar{\theta}(s_1),\bar{\theta}(s_2),\ldots,\bar{\theta}(s_k))$ which minimize D subject to the restriction that $\bar{\theta}$ is isotone. In the bioassay example, there also were the raw or preliminary estimates (the $\hat{p}_i$). However, as we shall see, they arise naturally from the objective function. For A a nonempty subset of S and Φ a real valued function on S, we denote by $D(\Phi|A)$ the portion of the value of the objective function at Φ which is attributable to the samples corresponding to points s_i with $s_i \in A$. In the bioassay example (recall $\sum_A = \sum_{\{i:\, s_i \in A\}}$),

$$D(p_1,p_2,\ldots,p_k|A) = -\textstyle\sum_A\{n_i\hat{p}_i \ln p_i + n_i(1-\hat{p}_i)\ln(1-p_i)\},$$

where we agree $0 \cdot \infty = 0$ and that $D(p_1,p_2,\ldots,p_k|A) = \infty$ if $p_i \notin [0,1]$ for some i with $s_i \in A$. For x a real number and A a nonempty subset of S, set

$$d(x|A) = D(\Phi|A) \text{ where } \Phi(s_i) = x \text{ for } i = 1,2,\ldots,k$$

and suppose that M(A) minimizes $d(x|A)$. Again returning to the bioassay problem, $d(p|A) = -(\sum_A n_i\hat{p}_i)\ln p - (\sum_A n_i(1-\hat{p}_i))\ln(1-p)$ for $p \in [0,1]$ and $M(A) = \sum_A n_i\hat{p}_i/\sum_A n_i$, which is the M considered in Section One. Furthermore $\hat{p}_i = M(\{s_i\})$. So in this general setting, the preliminary estimates are $\hat{\theta}(s_i) = M(\{s_i\})$, that is $\hat{\theta}(s_i)$ minimizes the part of $D(\Phi)$ attributable to the sample at s_i.

So in general, we are lead to the problem of minimizing $D(\Phi)$ subject to the restriction that Φ is a real valued function on S which is isotone with respect to some partial order. We expect that the solution will be expressed in terms of a set function M defined on the nonempty subsets of S with M(A) related to $D(\Phi|A)$, the restriction of $D(\Phi)$ to A. To make precise the fact that $D(\Phi|A)$ is to be the restriction of $D(\Phi)$ to A, we suppose that for each A, a nonempty subset of S, there is a function $D(\cdot|A)$ with domain the real valued functions for S and that these functions satisfy the following regularity conditions:

(5) $D(\Phi) \equiv D(\Phi|S)$ and if Φ and $\sum$ are functions on S with $\Phi(s_i) = \sum(s_i)$ for $s_i \in A$, then $D(\Phi|A) = D(\sum|A)$.

(6) if $A \subset B \subset S$ and if Φ and $\sum$ are functions on S with $\Phi(s_i) = \sum(s_i)$ for $s_i \in B-A$ and $D(\Phi|A) \leq D(\sum|A)$, then $D(\Phi|B) \leq D(\sum|B)$.

In the typical applications of this theory $D(\Phi|A)$ is either $\sum_A d(\Phi(s_i)|\{s_i\})$, $\Pi_A d(\Phi(s_i)|\{s_i\})$ or $\max_A d(\Phi(s_i)|\{s_i\})$ with $d(\cdot|A) \geq 0$ and in these cases (5) and (6) hold. In the bioassay example, $D(\Phi|A) = \sum_A d(\Phi(s_i)|\{s_i\})$ and $d(\cdot|A) \geq 0$ and so (5) and (6) hold in that case.

Before giving a solution to this restricted optimization problem, another definition is needed. A set function M defined on the nonempty subsets of S is a <u>Cauchy mean value function</u> provided $M(A) \wedge M(B) \leq M(A \cup B) \leq M(A) \vee M(B)$ for any nonempty disjoint subsets A and B. If in addition, both inequalities are strict whenever $M(A) \neq M(B)$, then M is said to be a <u>strict</u> Cauchy mean value function.

THEOREM 1. If the functions $D(\cdot|A)$ satisfy conditions (5) and (6), if M is a Cauchy mean value function and if $d(\cdot|A)$ is nonincreasing on $(-\infty, M(A)]$ and nondecreasing on $[M(A),\infty)$ for each nonempty subset A, then $\bar{\theta} = (\bar{\theta}(s_1), \bar{\theta}(s_2),\ldots,\bar{\theta}(s_k))$ with

(7) $$\bar{\theta}(s_i) = \max_{\{L':\ s_i \notin L'\}} \min_{\{L:\ s_i \in L\}} M(L-L')$$

$$= \min_{\{L:\ s_i \in L\}} \max_{\{L':\ s_i \notin L'\}} M(L-L')$$

minimizes $D(\Phi)$ subject to the restriction that Φ is isotone with respect to <<. Furthermore, if $\hat{\theta}$ is isotone then $\bar{\theta} = \hat{\theta}$.

Proof: See Robertson and Wright (1980).

(We will call the $\bar{\theta}(s_i)$ the isotonic estimates and will say that isotonizing M gives $\bar{\theta}$.)

It should be noted that the hypotheses of Theorem 1 are satisfied in the bioassay example. Conditions (5) and (6) have already been seen to hold. Examining the derivative of $d(x|A)$ for $0 < x < 1$, one sees that the assumption on $d(\cdot|A)$ is satisfied and so we only need to show that M is a Cauchy mean value function. But for any weighted average of the form $M(A) = \sum_A w(s_i)M(\{s_i\})/\sum_A w(s_i)$ with $w(s_i) > 0$ for $i = 1,2,\ldots,k$,

$$M(A \cup B) = \alpha M(A) + (1-\alpha)M(B) \text{ with } \alpha = \sum_A w(s_i)/\sum_{A\cup B} w(s_i)$$

for nonempty, disjoint subsets A and B. Hence, such an M is a strict Cauchy mean value function. In the bioassay example, $M(\{s_i\}) = \hat{p}_i$ and $M(A) = \sum_A n_i\hat{p}_i/\sum_A n_i$.

As was noted earlier, the max-min formula is typically not the most convenient method for computing these minimizing values.

THEOREM 2. If << is a total order on S and M is a Cauchy mean value function, then both formulas in (7) give the same values as the PAVA and the resulting values given by the PAVA do not depend on the order in which violators are pooled.

Proof. See the proof of Theorem 3.1 of Robertson and Waltman (1968).

THEOREM 3. If M is a strict Cauchy mean value function, then both formulas in (7) give the same values as the MLSA.

Proof. See the proof of Theorem 2.1 of Robertson and Wright (1980). Robertson and Wright also give a modified MLSA which is applicable for Cauchy mean value functions which are not necessarily strict.

It should be noted that although Theorem 1 does not apply if M is not a Cauchy mean value function, $\bar{\theta}(s_i)$ defined by the max-min formula in (7) for any set function M is isotone. This is true for both representations given but they are not necessarily equal unless M is a Cauchy mean value functions. We now show that the first representation is isotone, the proof of the other is similar. Suppose that $s_i << s_j$ and note that $\{L: s_i \varepsilon L\} \supset \{L: s_j \varepsilon L\}$ and $\{L': s_i \notin L'\} \subset \{L': s_j \notin L'\}$. Hence,

$$\bar{\theta}(s_i) \le \max_{\{L':\, s_i \notin L'\}} \min_{\{L:\, s_j \varepsilon L\}} M(L-L')$$

$$\le \max_{\{L':\, s_j \notin L'\}} \min_{\{L:\, s_j \varepsilon L\}} M(L-L') = \bar{\theta}(s_j).$$

In the next section we consider an order restricted inference problem which shows some of the pathology that arises when non Cauchy mean value functions are isotonized.

3. SOME ORDER RESTRICTED K-SAMPLE INFERENCES. In this section, we consider some of the properties of the isotonic estimates, $\bar{\theta}(s_i)$ for $i = 1,2,\ldots,k$, associated with certain choices of $D(\cdot)$. Particular attention will be given to a family of restricted optimization problems solved by Theorem 1. To motivate the objective functions of this family a one-sample problem is considered first.

Suppose that θ is a "typical value" of a measurement on some population, for instance θ may be a mean temperature, a median salary, etc. Suppose that

x_j, $j = 1,2,\ldots,n$, are sample values chosen from that population. One might estimate θ by choosing a value as "close" to the data as possible. In particular, one might choose $\hat{\theta}$ to minimize $\delta(x,\theta\cdot e)$ where $x = (x_1,x_2,\ldots,x_n)$, e is an n dimensional vector of ones, and δ is a metric on R_n. If the ℓ_r metrics were used, then $\hat{\theta}_r$ would be chosen to minimize

$$\delta_r(x,\theta\cdot e) = \begin{cases} \sum_{j=1}^n |x_j-\theta|^r & \text{if } 0 < r < 1 \\ (\sum_{j=1}^n |x_j-\theta|^r)^{1/r} & \text{if } 1 \leq r < \infty. \end{cases}$$

As $r \to 0$, $\delta_r(x,\theta\cdot e) \to \delta_0(x,\theta\cdot e) = \text{card}\{j: x_j \neq \theta, j=1,2,\ldots,n\}$ and as $r \to \infty$, $\delta_r(x,\theta\cdot e) \to \delta_\infty(x,\theta\cdot e) = \max_{1\leq j\leq n}|x_j-\theta|$. It is well known that $\hat{\theta}_0$ is the mode, a value in the sample which occurs with the greatest frequency (of course, $\hat{\theta}_0$ need not be unique), $\hat{\theta}_1$ is the median, the middle value when the sample is arranged in increasing order ($\hat{\theta}_1$ is not unique when n is even), $\hat{\theta}_2 = n^{-1}\sum_{j=1}^n x_j$ is the arithmetic mean, and $\hat{\theta}_\infty$ is the midrange, the average of the smallest and largest sample values. To be explicit, let us agree that $\hat{\theta}_1$ is the average of the two middle values if n is even. The estimates corresponding to values values of r other than $0,1,2,\infty$ are not commonly used, primarily because they are more difficult to compute. For $0 \leq r \leq 1$, the $\hat{\theta}_r$ need not be unique but for $1 < r \leq \infty$, they are. For $0 \leq r < 1$, the estimates are not Cauchy averages, that is the $\hat{\theta}_r$ value corresponding to a sample obtained by combining two samples is not necessarily between the $\hat{\theta}_r$ values for the individual samples, for $r = 1$ the estimate is a Cauchy average but not a strict one, and for $1 < r \leq \infty$ they are strict Cauchy averages. The arithmetic average is the most commonly used of these statistics. This is due, in part, to the fact that it is the m.l.e. for the mean of a normal distribution. Since extreme observations have considerable influence on $\hat{\theta}_2$ and $\hat{\theta}_1$ is very insensitive to extreme observations, one might consider using $\hat{\theta}_{r_0}$ for some $1 < r_0 < 2$ as an estimator of θ.

Returning to the k-sample ordered inference problems, we consider estimating typical values $\theta(s_i)$, $i = 1,2,\ldots,k$, which are known to be isotone with respect to a given partial order <<, by minimizing $D_r(\Phi)$, $1 \leq r \leq \infty$, subject to the restriction that Φ is isotone, where

$$D_r(\Phi|A) = (\sum_A \sum_{j=1}^{n_i} |x_{ij}-\Phi(s_i)|^r)^{1/r} \qquad \text{for } 1 \leq r < \infty,$$

and

$$D_\infty(\Phi|A) = \max_A \max_{j=1,2,\ldots,n_i} |x_{ij}-\Phi(s_i)|.$$

It is not difficult to show that the hypotheses of Theorem 1 are satisfied in these cases with $M_r(A)$ the ℓ_r-statistic which minimizes $d_r(x|A) = (\sum_A \sum_{j=1}^{n_i} |x_{ij}-x|^r)^{1/r}$ for $1 \leq r < \infty$ and $d_\infty(x|A) = \max_A \max_{j=1,2,\ldots,n_i} |x_{ij}-x|$. Hence, the max-min formula given in Theorem 1 yields the desired isotonic estimates, which we denote by $\bar{\theta}_r(s_i)$ for $1 \le r \le \infty$. However, Theorem 1 does not apply for $0 \le r < 1$.

The investigation of the properties of these estimators for small sample sizes by analytic means seems very difficult. But the results of a Monte Carlo study concerning their biases and mean squared errors will be discussed in one of the examples considered later. For fixed k, their large sample properties can be deduced from well-known properties of the preliminary estimators and the distance-reducing property given below which is due to Robertson and Wright (1974). This property applies not only to the estimates $\bar{\theta}_r$ but also to any $\bar{\theta}$ obtained from the max-min formula applied to any Cauchy mean value function M.

THEOREM 4. If M is a Cauchy mean value function and Φ is an isotone function on S, then

(8) $$\max_{1\le i\le k}|\bar{\theta}(s_i)-\Phi(s_i)| \le \max_{1\le i\le k}|\hat{\theta}(s_i)-\Phi(s_i)|.$$

Proof. Let s_j be a fixed element of S, let Φ be a fixed isotone function, and let $L_0 = \{s_\ell : \Phi(s_\ell) \le \Phi(s_j)\}$. Since L_0 is a lower layer,

$$\bar{\theta}(s_j)-\Phi(s_j) \le \max_{\{L':\ s_j \notin L'\}} M(L_0-L')-\Phi(s_j)$$

$$\le \max_{\{L':\ s_j \notin L'\}} \max_{L_0-L'} M(\{s_i\})-\Phi(s_j)$$

$$\le \max_{\{L':\ s_j \notin L'\}} \max_{L_0-L'} (M(\{s_i\})-\Phi(s_i))$$

$$\le \max_{1\le i\le k}|\hat{\theta}(s_i)-\Phi(s_i)|.$$

Similarly, it can be shown that $\bar{\theta}(s_j)-\Phi(s_j) \ge -\max_{1\le i\le k}|\hat{\theta}(s_i)-\Phi(s_i)|$ and the desired result follows since j was arbitrary.

This distance-reducing property applied to the bioassay problem shows that the maximum discrepancy between the isotonized estimate $\bar{p}_i$ and the true value p_i is no larger than the maximum discrepancy between the preliminary estimate $\hat{p}_i$ and the true value p_i, assuming the p_i are isotone. Other distance-reducing properties have been observed earlier. One is given in Theorem 1.1 of Barlow et al. (1972).

It follows immediately from Theorem 4 that the isotonization of Cauchy mean value functions preserves consistency in the case in which k is fixed.

THEOREM 5. If k is fixed, M is a Cauchy mean value function, θ is isotone, and $\hat{\theta}(s_i)$ is consistent for $\theta(s_i)$ for $i = 1,2,\ldots,k$, then

$$\max_{1\le i\le k}|\bar{\theta}(s_i)-\theta(s_i)| \to 0 \text{ as } \min_{1\le i\le k} n_i \to \infty.$$

For the bioassay example, it is well known that the relative frequencies $\hat{p}_i$ converge to the true probabilities p_i as $n_i \to \infty$ and so it follows that the isotonized estimates $\bar{p}_i$ are consistent for the p_i provided the p_i are isotone.

Applications of Theorem 5 to the $\overline{\theta}_r$ will be discussed in the examples later in this section.

Next we consider the asymptotic distribution of these isotonized estimates for the case when k is fixed. Again a general result is given and its application to the $\overline{\theta}_r$ will be discussed in the examples which follow.

THEOREM 6. Suppose that k is fixed, M is a Cauchy mean value function, $\theta(s_i)<\theta(s_j)$ whenever $s_i \ll s_j$ with $i \neq j$, and that there is a nondedreasing function g with $g(x) \to \infty$ as $x \to \infty$ for which $g(n_i)(\hat{\theta}(s_i) - \theta(s_i))$ has a limiting distribution with cumulative distribution function (c.d.f.) $H_i(x)$ for $i = 1,2,\ldots,k$. If each $n_i \to \infty$ with $g(\max_{1\le i\le k} n_i)/g(\min_{1\le i\le k} n_i)$ bounded then $g(n_i)(\overline{\theta}(s_i) - \overline{\theta}(s_i))$ has limiting distribution with c.d.f. $H_i(x)$ for $i = 1,2,\ldots,k$.

Proof. Choose $\varepsilon > 0$ so that $\theta(s_j) - \theta(s_i) \ge \varepsilon$ if $s_i \ll s_j$ and $i \neq j$, and note that if $\max_{1\le i\le k}|\hat{\theta}(s_i) - \theta(s_i)| \le \varepsilon/2$, then $\hat{\theta}(s_i) \le \theta(s_i) + \varepsilon/2 \le \theta(s_j) - \varepsilon/2 \le \hat{\theta}(s_j)$ provided $s_i \ll s_j$. Hence, if A denotes the set of sample values for which $\max_{1\le i\le k}|\hat{\theta}(s_i) - \theta(s_i)| \le \varepsilon/2$, then on $A, \hat{\theta}$ is isotone and thus $\overline{\theta} = \hat{\theta}$. So

$$g(n_i)(\overline{\theta}(s_i) - \theta(s_i)) = g(n_i)(\hat{\theta}(s_i) - \theta(s_i))I_A + g(n_i)(\overline{\theta}(s_i) - \theta(s_i))I_{A^c}$$
$$= g(n_i)(\hat{\theta}(s_i) - \theta(s_i)) + g(n_i)(\overline{\theta}(s_i) - \hat{\theta}(s_i))I_{A^c}.$$

The first term of the last expression has the desired limiting distribution. In considering the second term, note that $\hat{\theta}(s_i) - \theta(s_i) \to 0$ (in probability) since $g(n_i) \to \infty$ as $n_i \to \infty$ and so $I_{A^c} \to 0$ (in probability). But $g(n_i)|\hat{\theta}(s_i) - \overline{\theta}(s_i)| \le g(n_i)(|\hat{\theta}(s_i) - \theta(s_i)| + |\overline{\theta}(s_i) - \theta(s_i)|) \le 2g(\max_{1\le i\le k} n_i)\max_{1\le i\le k} |\hat{\theta}(s_i) - \theta(s_i)| \le 2(g(\max_{1\le i\le k} n_i)/g(\min_{1\le i\le k} n_i))\max_{1\le i\le k} g(n_i)|\hat{\theta}(s_i) - \theta(s_i)|$, which has a limiting distribution. So the second term converges (in probability) to zero and the proof is completed.

In the bioassay example, $\sqrt{n_i}(\hat{p}_i - p_i)$ has a normal limiting distribution with mean 0 and variance $p_i(1-p_i)$. So if the order restrictions imposed on the p_i are strict inequalities and the $n_i \to \infty$ with $\max_{1\le i\le k} n_i/\min_{1\le i\le k} n_i$ bounded then $\sqrt{n_i}(\overline{p}_i - p_i)/(p_i(1-p_i))^{1/2}$ and $\sqrt{n_i}(\overline{p}_i - p_i)/(\overline{p}_i(1-\overline{p}_i))^{1/2}$ both have standard normal limiting distributions. So asymptotically $\overline{\theta}(s_i)$ and $\hat{\theta}(s_i)$ behave alike, but the $\overline{\theta}(s_i)$ are isotone and are more accurate for small sample sizes.

EXAMPLE. (ℓ_2 or least squares problem) The ordered ℓ_2 optimization problem and related statistical problems have received considerable attention. Barlow, et al. (1972) gives a thorough discussion of this problem and the related inferences including both estimation procedures and tests of hypotheses. To motivate such inferences we consider the problem of estimating the mean score on an examination which measures reading skills. The same examination is given to third, fourth, fifth and sixth grade students and the mean scores for a certain group of students, say minority students in a given school district, are denoted by $\theta(3)$, $\theta(4)$, $\theta(5)$ and $\theta(6)$. One could use the average of the scores in the sample from grade i to estimate $\theta(i)$ for $i = 3,4,5,6$. However, if the sample sizes are not large violations do occur in practice, that is the average score of fourth grade students may exceed that for fifth grade students, of course, it is believed that $\theta(3) < \theta(4) < \theta(5) < \theta(6)$ and ad hoc mean are sometimes used to "smooth" these averages to obtain increasing estimates. As we shall see, if the scores are normally distributed and the m.l.e.s subject to the restriction that the $\theta(i)$ are nondecreasing are desired then we are led to an ordered ℓ_2 optimization problem. However, the ordered ℓ_2 estimates are not completely satisfactory since in the case of a violation the estimates are equal. A means for overcoming this difficulty is discussed later.

If the $\theta(s_i)$ are means and the x_{ij} are normally distributed about these means with common variances σ^2, then the order restricted m.l.e.s of the $\theta(s_i)$ are obtained by maximizing

$$(2\pi\sigma^2)^{-\sum_{i=1}^{k} n_i/2} \exp\{-(2\sigma^2)^{-1}\textstyle\sum_{i=1}^{k}\sum_{j=1}^{n_i}(x_{ij}-\Phi(s_i))^2\}$$

subject to the restriction that Φ is isotone or equivalently, by minimizing

$$\textstyle\sum_{i=1}^{k}\sum_{j=1}^{n_i}(x_{ij}-\Phi(s_i))^2$$

subject to the restriction that Φ is isotone. As we have seen, a solution to this restricted optimization problem is the $\overline{\theta}$ given by the max-min formula (one can use the PAVA if the order relation is total or the MLSA for partial orders) applied to $M_2(A) = \sum_A\sum_{j=1}^{n_i} x_{ij}/\sum_A n_i$. It is noted in Theorem 1.1 of Barlow, et al. (1972) that the solution is unique in the ℓ_2 case. Brunk (1955) obtained these estimates in a more general setting which allowed the variances to depend on i. (Theorem 1 can also be used to obtain the solution to this more general problem by choosing M to be a weighted average with appropriately chosen weights.) The preliminary estimates $\hat{\theta}_2(s_i) = M(\{s_i\}) = \sum_{j=1}^{n_i} x_{ij}/n_i$ are the sample means, typically denoted by $\overline{x}_i$, and the $\overline{\theta}_2(s_i)$ are referred to as isotonized means.

We now give a brief summary of a Monte Carlo study of the biases and mean squared errors (m.s.e.s) of the isotonic estimates and the m.l.e.s based on the assumption that $\theta(s)$ is linear. With $k = 9$ and $s_i = i/10$, independent random samples of $n_i = 4$ normally distributed random variables with mean $\theta(s_i)$ and variance 1 were simulated for $i = 1,2,\ldots,9$. The choices for $\theta(s)$ were $4s$, $4s^2$, $4\sqrt{s}$, $2(2s-1)^5$ and $2(2s-1)^{1/5}$. The first function was included to see how the isotonic estimates performed if in fact the $\theta(s_i)$ were linear and the other functions have in various patterns and to varying degrees intervals of slow increase and rapid increase. Then for each of the functions and $i = 1,2,\ldots,9$, the differences $\overline{\theta}_2(s_i)-\theta(s_i)$ were computed. Based on 10,000 iterations of this procedure the average of the $\overline{\theta}_2(s_i)-\theta(s_i)$ values was computed as an estimate of the bias of the isotonic estimate at s_i if the true means were $\theta(s_i)$ and the average of the $(\bar{\theta}_2(s_i)-\theta(s_i))^2$ values was computed as an estimate of the m.s.e. of the isotonic estimate at s_i if the true means were $\theta(s_i)$. For a given function θ, the estimated m.s.e.s at s_i was summed for $i = 1,2,\ldots,9$ to give an estimate of the total m.s.e. of $\bar{\theta}_2$ for that choice of θ. This process was repeated for $n_i = 12$ and then for $n_i = 4$ and 12 it was carried out again for $\hat{\theta}_\ell$, the m.l.e. based on the assumption that θ is linear.

The isotonic estimator has a positive bias at points near the upper end point of the interval and has a negative bias at points near the lower end point of the interval. This is because $\bar{\theta}_2(s_i)$ is forced upwards because of the isotonic assumption if i is large and is forced downwards if i is small. (See Cryer, et al. (1972) for a further discussion of bias.) Some of the main results of the m.s.e. study are given in Table 1, a detailed discussion is contained in Wright (1978).

Table 1: Total Mean Squared Error for $\hat{\theta}_\ell$ and $\bar{\theta}_2$.

	$n_i \equiv 4$		Functions	$n_i \equiv 12$		
Estimator	$4s$	$4s^2$	$2(2s-1)^{1/5}$	$4s$	$4s^2$	$2(2s-1)^{1/5}$
$\hat{\theta}_\ell$	.50	.98	2.84	.17	.66	2.52
$\bar{\theta}_2$	1.36	1.34	1.35	.58	.56	.49

As we expected $\hat{\theta}_\ell$ had smaller m.s.e. than $\bar{\theta}_2$ if θ is linear and the reverse is true if θ is highly nonlinear. For the function $4s^2$, $\hat{\theta}_\ell$ had the smaller m.s.e. for small n_i but as n_i became larger this was reversed. It should be noted that the total m.s.e. for the preliminary estimator is the same no matter what $\theta(s)$ is. For $n_i \equiv 4$ the total m.s.e. of $\hat{\theta}_2$ is 2.25 and for $n_i \equiv 12$ it is .75. R. V. Hogg proposed combining these two estimators $\hat{\theta}_\ell$ and $\bar{\theta}_2$ to obtain an estimator with smaller m.s.e. in the nearly linear cases without losing the protection that $\bar{\theta}_2$ provides in the nonlinear case. So in the Monte Carlo study being discussed a third estimator $\theta_c(s) = \gamma\hat{\theta}_\ell(s) + (1-\gamma)\bar{\theta}_2(s)$ was considered with $\gamma = (k-2)R/\sum_{i=1}^k n_i$ and $R = \sum_{i=1}^k \sum_{j=1}^{n_i}(x_{ij}-\bar{\theta}(s_i))^2/\sum_{i=1}^k \sum_{j=1}^{n_i}(x_{ij}-\hat{\theta}_\ell(s_i))^2$. (A rationale for this choice of λ is given in the paper being discussed.) For all the cases in this Monte Carlo study the total m.s.e. of θ_c was never larger than that of $\bar{\theta}_2$ and in the linear case it was 20 percent smaller than that of $\bar{\theta}_2$ if $n_i \equiv 4$ and 25 percent smaller if $n_i \equiv 12$. The estimator θ_c also might be preferred over $\bar{\theta}_2$ since for almost all the samples studied it was a strictly increasing function while $\bar{\theta}_2$ often had the same value for two or more elements in S. We saw in our discussion concerning the estimation of the mean scores on the reading skills examination that strictly increasing estimates are desirable in some situations.

Next we consider the large sample properties of $\bar{\theta}_2$. We no longer need to assume that the x_{ij} are normally distributed but only that $x_{ij}-\theta(s_i)$ are independent and identically distributed with mean zero and variance $\sigma^2 \varepsilon (0,\infty)$. Under these conditions $\bar{x}_i$ is consistent for $\theta(s_i)$ and $\sqrt{n_i}(\bar{x}_i - \theta(s_i))/\sigma$ has a standard normal limiting distribution. So $\bar{\theta}_2(s_i)$ is consistent for $\theta(s_i)$ for $i = 1,2,\ldots,k$ provided θ is isotone. If $\theta(s_i) < \theta(s_j)$ whenever $s_i << s_j$ with $i \neq j$ and if the $n_i \to \infty$ with $\max_{1 \leq i \leq k} n_i / \min_{1 \leq i \leq k} n_i$ bounded then $\sqrt{n_i}(\bar{\theta}_2(s_i) - \theta(s_i))/\sigma$ has a standard normal limiting distribution. This statement remains valid if σ is replaced by $\hat{\sigma} = (\sum_{i=1}^k \sum_{j=1}^{n_i} (x_{ij} - \bar{\theta}_2(s_i))^2 / \sum_{i=1}^k n_i)^{1/2}$.

EXAMPLE. (ℓ_1 or least absolute deviations problem) Suppose that we wish to minimize

$$\sum_{i=1}^k \sum_{j=1}^{n_i} |x_{ij} - \Phi(s_i)| \quad \text{subject to the restriction that } \Phi \text{ is isotone.}$$

We have already seen that a solution to this restricted optimization problem is given by the max-min formula applied to the set function M_1, where $M_1(A)$ is the median of the x_{ij} for $j = 1,2,\ldots,n_i$ and all i for which $s_i \varepsilon A$. Recall the solution is not necessarily unique even in the case $k = 1$. The preliminary estimates are $\hat{\theta}_1(s_i) = M(\{s_i\})$ which is the median of the ith sample. The $\bar{\theta}_1(s_i)$ are referred to as isotonized medians. The restricted ℓ_1 problem has been considered by Robertson and Waltman (1968), Brunk and Johansen (1970), Cryer, et al. (1972), Robertson and Wright (1973), and Casady and Cryer (1976). The problem arises when obtaining the m.l.e.s of the location parameters of k bilateral exponential distributions and these parameters are subject to an isotonic constraint. A Monte Carlo study of the biases and m.s.e.s of $\bar{\theta}_1$ as well as a comparison of $\bar{\theta}_1$ and $\bar{\theta}_2$ are given in Cryer, et al. (1972).

To study the large sample properties of the $\bar{\theta}_1(s_i)$, we assume that the $x_{ij} - \theta(s_i)$ are independent and identically distributed with c.d.f. F. We assume that F has median zero and that F has continuous derivative f in a neighborhood of zero with $f(0) > 0$. Then $\sqrt{n_i}(\hat{\theta}_1(s_i) - \theta(s_i))$ has a normal limiting distribution with mean zero and variance $(4f^2(0))^{-1}$. Hence, $\bar{\theta}_1(s_i)$ is consistent for $\theta(s_i)$ provided θ is isotone. If $\theta(s_i) < \theta(s_j)$ whenever $s_i << s_j$ with $i \neq j$ and if the $n_i \to \infty$ with $\max_{1 \leq i \leq k} n_i / \min_{1 \leq i \leq k} n_i$ bounded then $\sqrt{n_i}(\bar{\theta}_1(s_i) - \theta(s_i))$ has a normal limiting distribution with mean zero and variance $(4f^2(0))^{-1}$.

Similar results hold for percentiles other than the median. (See Casady and Cryer (1976).) The 100p th percentile of the sample $x_1, x_2, \ldots, x_n$ minimizes $\sum_{i=1}^n g(x_i, \theta)$ where $g(x,\theta) = p(x-\theta)$ if $x \geq \theta$ and $g(x,\theta) = (1-p)(\theta - x)$ if $x < \theta$. So isotonized percentiles minimize

$$\sum_{i=1}^k \sum_{j=1}^{n_i} g(x_{ij}, \Phi(s_i)) \quad \text{subject to the restriction that } \Phi \text{ is isotone.}$$

EXAMPLE. (ℓ_∞ or least maximum deviations problem) If we desire to minimize

$$\max\{|x_{ij} - \Phi(s_i)| : j = 1,2,\ldots,n_i,\ i = 1,2,\ldots,k\} \tag{9}$$

subject to the restriction that Φ is isotone, then a solution is given by the max-min formula applied to the set function M_∞, where $M_\infty(A)$ is the midrange of the x_{ij} for $j = 1,2,\ldots,n_i$ and all i for which $s_i \in A$. The preliminary estimates are $\hat{\theta}_\infty(s_i) = M_\infty(\{s_i\})$ which is the midrange of the ith sample. The $\bar{\theta}_\infty(s_i)$ are referred to as isotonized midranges. It is interesting to note that the one-sample ℓ_∞ problem has a unique solution, but this is not so for $k > 1$. For example, if $k = 2$, $s_1 << s_2$, $n_1 = 2$, $n_2 = 1$, $x_{11} = 0$, $x_{12} = 1$ and $x_{21} = 1/4$, then $\Phi(s_1) = 1/2$ and $\Phi(s_2) = x$ with $1/2 \leq x \leq 3/4$ minimizes (9). The solution with $x = 1/2$ is the one we have denoted by $\bar{\theta}_\infty$. The ℓ_∞ problem has been studied by Ubhaya (1974a,b) and Barlow and Ubhaya (1971). They have characterized all the solutions while Theorem 1 only gives one.

The asymptotic results for midranges given in Sarhan and Greenberg (1962, p. 81) may be combined with Theorems 5 and 6 to obtain large sample results for $\bar{\theta}_\infty$.

EXAMPLE. (ℓ_0 problem) Consider the problem of minimizing

(10) $D_0(\Phi) = \sum_{i=1}^{k}\sum_{j=1}^{n_i} I_{[x_{ij} \neq \Phi(s_i)]}$ subject to the restriction that Φ is isotone. As we have seen $M_0(A)$ is a mode of the observations x_{ij} for $j = 1,2,\ldots,n_i$ and all i for which $s_i \in A$ and M_0 is not a Cauchy mean value function. We now give an example to show that the max-min formula applied to M_0 does not necessarily minimize (10). Let $k = 2$, $s_1 << s_2$, $n_1 = n_2 = 5$, $x_{ij} = 0$ for $i,j = 1,2$, $x_{1j} = 1$ for $j = 3,4,5$ and $x_{2j} = 2$ for $j = 3,4,5$. The max-min formula gives $\bar{\theta}_0(s_1) = 0$ and $\bar{\theta}_0(s_2) = 2$ but $D(\Phi) < D(\bar{\theta}_0)$ if $\Phi(s_1) = 1$ and $\Phi(s_2) = 2$. The MLSA gives $\Phi_0(s_1) = \Phi_0(s_2) = 0$ and $D(\Phi) < D(\Phi_0)$. The PAVA gives the minimizing value for this data set but if the samples at s_1 and s_2 are interchanged this is not the case. The fact that M is not a Cauchy mean value function also affects the large sample properties of $\bar{\theta}_0$. See Robertson and Wright (1974).

4. ISOTONIC REGRESSION. In the preceeding section, inferences for k ordered parameters were discussed. The parameters were also thought of as an isotone function on the finite set S. In a typical regression setting, functions defined on a rectangle in some Euclidean space are considered. For instance, in the bioassay example the dosage level s might be the concentration of the drug in a serum expressed as a proportion. The probability of response R could be thought of as a function p(s) defined on a subinterval of [0,1] and it might be reasonable to assume that p is a nondecreasing function, that is that p is an isotone function with respect to the usual order on the reals.

As another example, one might want to estimate $\theta(s_1,s_2)$, the mean grade point average at the end of the first year of college for those students whose high school percentile rank was s_1 and their percentile rank on a certain entrance exam was s_2. So we wish to estimate $\theta(s)$ where $s = (s_1,s_2)$ under the assumption that θ is isotone with respect to the usual partial order on R_2.

In general, let $\Delta = X_{i=1}^{\beta}(a_i,b_i)$ with $-\infty \leq a_i < b_i \leq \infty$ for $i = 1,2,\ldots,\beta$, suppose that for each $s \in \Delta$ there is a distribution with parameter $\theta(s)$ and suppose that the function θ defined on Δ is isotone with respect to the usual partial order on R_β. The parameters might be probabilities, means, medians, etc., but to illustrate the ideas involved we suppose they are means. If θ is to be estimated, then observation points t_j for $j = 1,2,\ldots,n$ are selected in Δ and observations on each of the distributions associated with these observations points are taken. Let x_j denote the observation taken at t_j for $i = 1,2,\ldots,n$ and let $s_1,s_2,\ldots,s_k$ denote the distinct observation points in $t_1,t_2,\ldots,t_n$ (k depends on n). Label the observations at s_i, x_{ij} for $j = 1,2,\ldots,n_i$ and $i = 1,2,\ldots,k$, of course, $\sum_{i=1}^{k} n_i = n$. Estimate θ at an observation point by

$$\bar{\theta}(s_i) = \max_{\{L':\ s_i \notin L'\}} \min_{\{L:\ s_i \in L\}} M_2(L-L')$$

and let $\bar{\theta}$ be an isotone extension to Δ. There are numerous ways to extend $\bar{\theta}$ from $\{s_1,s_2,\ldots,s_k\}$ to Δ. One such way is to set $\bar{\theta}(s) = \min\{\bar{\theta}(s_j): s \ll s_j\}$ and if there is no $s_j \gg s$, then set $\bar{\theta}(s) = \max\{\bar{\theta}(s_j): j = 1,2,\ldots,n\}$. This is an extension because the $\bar{\theta}(s_i)$ are isotone. If $s,t \in \Delta$ with $s \ll t$, then $\{s_j: s \ll s_j\} \supset \{s_j: t \ll s_j\}$ and considering separately the cases $\{s_j: t \ll s_j\}$ is empty or nonempty, we see that $\bar{\theta}(s) \ll \bar{\theta}(t)$.

The small sample properties of $\bar{\theta}(s)$ at the observation points were examined in the ℓ_2 example in the last section and the behavior of $\bar{\theta}(s)$ for an s which is not an observation point depends on the manner in which $\bar{\theta}$ has been extended from $S = \{s_1,s_2,\ldots,s_k\}$ to Δ. The properties of $\bar{\theta}(s)$ for large n are not as easy to obtain in this setting as they were in the k-sample problems. If $\bar{\theta}$ is to be consistent for a strictly increasing function θ defined on Δ, it is intuitively clear that $\{t_1,t_2,\ldots,t_n\}$ must become dense in Δ. For if there is a rectangle in Δ which contains none of the t_n, then we can not determine the behavior of θ on this rectangle. In Hanson et al. (1973) it is argued that if $\bar{\theta}$ is to be strongly consistent for θ then the following must hold:

(11) $\liminf_{n\to\infty} \text{card.}\{1 \leq j \leq n:\ t_j \in J\}/n > 0$

for each nondegenerate rectangle $J \subset \Delta$. So in this regression setting we consider sequences of observation points for which the number of observations at any particular point may be small but they become dense in Δ in a manner so that (11) is satisfied. If $\beta = 1$ then no further assumption on the t_k is needed for $\bar{\theta}$ to be consistent for θ, but for $\beta > 1$ this is not so. Any easy method of obtaining observation points which satisfy the needed conditions is to generate them randomly according to a distribution with support Δ. (Wright (1979) gives a sufficient condition for nonrandom observation points.)

We now consider random observation points. Suppose that $x_1, t_1, x_2, t_2, \ldots$ are independent, the t_k all have the same distribution with c.d.f.F, and the $x_k - \theta(t_k)$ are identically distributed with mean zero. Assume that F assigns positive probability to each nondegenerate rectangle in Δ, so that, with probability one, the t_k satisfy (11) and assume that θ is continuous. In the case $\beta = 1$, Brunk (1970) has shown that $\bar{\theta}$ is strongly consistent for θ. (Brunk did not assume that the $x_k - \theta(t_k)$ are identically distributed but imposed more stringent moment restrictions. Hanson et al. (1973) considered a different condition which reduces to the one given here in the identically distributed case.) For the case $\beta > 1$, Wright (1979) has shown that $\bar{\theta}$ is strongly consistent for θ if the distribution associated with F is absolutely continuous (with respect to Lebesgue measure), that is if F has a probability density function (p.d.f.). Consistency results for isotone regression estimates determined by the max-min formula for functions other than the mean are considered in Robertson and Wright (1975).

For the case $\beta = 1$, we consider the asymptotic distribution of $\bar{\theta}$ at a fixed point s_0. We assume that the differences $x_k - \theta(t_k)$ have a finite, positive variance σ^2, that $\theta'(s_0) > 0$ and that F has a continuous derivative in a neighborhood of s_0, with $F'(s_0) > 0$. Brunk (1970) obtained the asymptotic distribution of

$$\left(\frac{2F'(s_0)n}{\sigma^2\theta'(s_0)}\right)^{1/3}(\bar{\theta}(s_0) - \theta(s_0)).$$

It is the same as the distribution of the slope at zero of the greatest convex minorant of $W(t) + t^2$, where $W(t)$ is the standard two-sided Wiener-Levy process. This distribution has p.d.f. $\psi(x/2)/2$ where $\psi(x) = u_1(x^2,x)u_1(x^2,-x)/2$ and $u(x,t)$ is the solution of the heat equation $u_{11}(x,t) = 2u_2(x,t)$ for $x < t^2$, subject to the boundary conditions $u(x,t) = 1$ for $x \geq t^2$ and $u(x,t) \to 0$ as $x \to -\infty$. (Brunk considered equally spaced observation points. The modification for other observation points was considered by Leurgans (1978) and Wright (1981). In both of these papers the assumption $\theta'(s_0) > 0$ is also relaxed.)

Recall that in the k-sample problems for proportions, means and medians the norming sequences for the limiting distribution results were proportional to $n^{1/2}$ but in the regression case they are proportional to $n^{1/3}$. Hence, the rate of convergence of $\bar{\theta}(s_0)$ to $\theta(s_0)$ is of order $n^{-1/3}$ whereas in the k-sample problem the order is $n^{-1/2}$. Of course, we would not expect the estimation of a function to be as efficient as the estimation of the value of the function at a few points (unless the functional form were known). If this result concerning the limiting distribution of $\theta(s_0)$ is to be used in determining large sample confidence intervals or tests of hypotheses, then the percentiles of this limit distribution are needed. Furthermore, asymptotic results for $\beta > 1$

and for choices of M other than the mean would be useful.

5. OTHER APPLICATIONS. In Chapter One of Barlow, et al. (1972) the ideas discussed here (order restricted optimization and ordered estimation) are applied to problems in multidimensional scaling, production scheduling and inventory planning. In statistical settings these ideas have proved useful in testing hypotheses about ordered parameters, estimating stochastically ordered distributions and making inferences for monotone failure rates. See Chapters 3, 4, 5 and 6 of Barlow, et al. (1972).

6. ACKNOWLEDGMENT. This work was partially sponsored by the National National Science Foundation under Grants MCS 75-23576 and MCS 78-01514.

BIBLIOGRAPHY

1. Ayer, M., Brunk, H. D., Ewing, G. M., Reid, W. T. and Silverman, E. (1955). An empirical distribution function for sampling with incomplete information. Ann. Math. Statist. 26, 641-647.

2. Barlow, R. E., Bartholomew, D. J., Bremner, J. M. and Brunk, H. D. (1972). Statistical Inference Under Order Restrictions, New York: Wiley.

3. Barlow, R. E. and Ubhaya, V. A. (1971). "Isotonic Approximation" in Proceedings of the Symposium on Optimizing Methods in Statistics. New York: Academic Press, Inc.

4. Brunk, H. D. (1955). Maximum likelihood estimates of ordered parameters. Ann. Math. Statist. 26, 607-616.

5. Brunk, H. D. (1958). On the estimation of parameters restricted by inequalities. Ann. Math. Statist. 29, 437-454.

6. Brunk, H. D. (1965). Conditional expectation given a σ-lattice and applications. Ann. Math. Statist. 36, 1339-1350.

7. Brunk, H. D. (1970). "Estimation of Isotonic Regression" in Nonparametric Techniques in Statistical Inference. Cambridge: Cambridge University Press, 177-195.

8. Brunk, H. D. and Johansen, S. (1970). A generalized Radon-Nikodym derivative. Pacific J. Math. 34, 585-617.

9. Casady, Robert J. and Cryer, J. D. (1976). Monotone percentile regression. Ann. Statist. 4, 532-541.

10. Cryer, J. D., Robertson, Tim, Wright, F. T. and Casady, Robert J. (1972). Monotone median regression. Ann. of Math. Statist. 43, 1459-1469.

11. Hanson, D. L., Pledger, Gordon and Wright, F. T. (1973). On consistency in monotone regression. Ann. Statist. 1, 401-421.

12. Leurgans, Sue E. (1978). Asymptotic distribution theory in generalized isotonic regression. Ph.D. dissertation, Stanford University.

13. Robertson, Tim and Waltman, Paul (1968). On estimating monotone parameters. Ann. Math. Statist. 39, 1030-1039.

14. Robertson, Tim and Wright, F. T. (1973). Multiple isotonic median regression. Ann. Statist. 1, 422-432.

15. Robertson, Tim and Wright, F. T. (1974). A norm reducing property for isotonized Cauchy mean value functions. Ann. Statist. 2, 1302-1307.

16. Robertson, Tim and Wright, F. T. (1975). Consistency in generalized isotonic regression, Ann. Statist. 3, 350-362.

17. Robertson, Tim and Wright, F. T. (1980). Algorithms in order restricted statistical inferences and the Cauchy mean value property. Ann. Statist. 8, (to appear).

18. Sarhan, A. E. and Greenberg, B. G. (1962). Contributions to Order Statistics. New York: Wiley.

19. Ubhaya, Vasant A. (1974). Isotone Optimization I. J. Approximation Theory 12, 146-159.

20. Ubhaya, Vasant A. (1974). Isotone Optimization II. J. Approximation Theory 12, 315-331.

21. Van Eeden, C. (1956). Maximum likelihood estimation of ordered probabilities. Indag. Math. 18, 444-455.

22. Van Eeden, C. (1957). Maximum likelihood estmation of partially or completely ordered parameters, I and II. Indag. Math. 19, 128-136, 201-211.

23. Wright, F. T. (1978). Estimating strictly increasing regression functions. J. Amer. Statist. Assoc. 73, 636-639.

24. Wright, F. T. (1979). A strong law for variables indexed by a partially ordered set with applications to isotone regression. Ann. Probability 7, 109-127.

25. Wright, F. T. (1981). The asymptotic behavior of monotone regression estimates. Ann. Statist. 9, (to appear).

DEPARTMENT OF STATISTICS
UNIVERSITY OF IOWA
IOWA CITY, IOWA 52242

DEPARTMENT OF MATHEMATICS
UNIVERSITY OF MISSOURI-ROLLA
ROLLA, MO 65401

Proceedings of Symposia in Applied Mathematics
Volume 23
1980

TIME SERIES: MODEL ESTIMATION, DATA ANALYSIS AND ROBUST PROCEDURES

R. Douglas Martin, University of Washington

1. INTRODUCTION

A time series data set is simply a collection of measurements of observed values $x_{t_1}, x_{t_2}, \ldots, x_{t_n}$ of some phenomenon which are indexed by a set of measurement times $t_1 < t_2 < \ldots < t_n$. Thus a time series is a _structured_ data set whose structure is specified by a measurement-time parameter t. At the risk of being too formal for the applied scientist, those who write about time series often like to denote a time series by $\{x_t, t \in \mathcal{T}\}$ where $\mathcal{T}$ is an index set of allowable values of t. Since the formalism is often convenient, we yield to the temptation to use it.

A number of distinctly different types of measurement times occur in practice. One of the most commonly encountered types is that of _equidistant spacing_: $t_i - t_{i-1} = \tau$, $2 \le i \le n$. In this case the index set is $\mathcal{T} = \{t_i = t_1 + (i-1)\tau, 1 \le i \le n\}$. However, it is common practice in this case to relabel the series as $x_1, x_2, \ldots, x_n$, then taking $\mathcal{T} = \{1, 2, \ldots, n\}$ for convenience and remembering that the time interval between observations is τ. Such time series are usually called _discrete-parameter time series_. For econometric modeling problems we often take _$\tau = 1$ for monthly data_ so that quarterly data corresponds to _$\tau = 4$_ and annual data corresponds to _$\tau = 12$_.

In engineering and physical sciences problems equispaced data often arises as a result of _controlled uniform sampling_ of a _continuous time-parameter series_ $\{x_t, t \in \mathcal{T}\}$ with $\mathcal{T} = [0,T]$, i.e., t is allowed to take on all real values in the closed interval $[0,T]$. The sampled series is $\tilde{x}_i =$

1980 Mathematics Subject Classification 62M10, 62F35.

$x_t|_{t=i\Delta}$, $1 \leq i \leq n$ where the sampling interval is Δ and T is chosen so that $T = n\Delta$. In this case a problem facing the experimenter is that of choosing a sufficiently high sampling rate $f_s = 1/\Delta$ so that essentially no information contained in x_t is lost by recording only the uniformly spaced samples $\tilde{x}_i$, $1 \leq i \leq n$. This issue is discussed in Sections 2.5 and 8.7 of Bloomfield (1976).

Another important class of measurement times is that associated with continuous time-parameter processes where observations are available only at randomly-spaced observation times $t_1 < t_2 < \ldots < t_n$ in the interval [0,T]. The experimenter may or may not have some control over the probabilistic specification of the observation times. An important example where the experimenter knows the probabilistic specification of the observation times (but has no control over them) occurs in jet-engine noise measurement experiments. The jet-flow velocity is measured by "seeding" the air intake with particulates whose velocities at a fixed point are measured using laser techniques. The arrival times of the particulates at the measurement point are random, but the randomness characteristics are known with high accuracy.

There is also a quite distinctive class of time series where the measurements or observations themselves are the occurrence times of events, such as the arrival times of messages for transmittal over a communications link or the arrivals of automobiles at a busy intersection. In such situations the focus of attention is on the event (or arrival) times themselves rather than on a value (e.g., an amplitude, etc.) associated with the event time, as in the jet-noise measurement experiment. Time series of this type are sometimes called point processes, and the associated continuous parameter time series x_t defined as the number of occurrences of events in the interval [0,t] is termed a counting process. Although such processes are quite important in view of their frequent occurrence in various areas of engineering and applied science, we shall not deal with them further here (see Bartlett, 1978, Sec. 3.42 and Cox and Lewis, 1966, Chap. 4).

In these brief notes we shall limit our attention solely to equi-spaced discrete-parameter time series. This case has received a great deal of attention because it arises frequently in practice and has a tractable theory.

Qualitative Features. Even though we are restricting attention to discrete time series data $x_1, x_2, \ldots, x_n$ (with sampling interval τ suppressed), such data exhibits a wide range of qualitative behavior which depends upon the source of the data. Time series often exhibit various forms of non-homogeneous or non-stationary behavior (stationarity will be defined shortly). Some possibilities are displayed in Figures 1a, 1b and 2. The oceanographic data in Figure 1a displays trendy behavior without obvious additional structure, wherease the RESX data of Figure 1b exhibits a fairly linear trend with a growing seasonal variation superimposed. Weekly male sperm count data is displayed in Figure 2.

An example of millimeter waveguide diameter measurements is given in Figure 4 of Kleiner, Martin and Thomson (1979). The latter time series is quite "smooth" and the sample size n = 1024 is large compared with the other data sets. Typically, but not always, one gets much larger sample sizes in engineering and applied sciences problems than in economics and biomedical problems, and this distinction often has a considerable effect in determining how narrowly one focuses on purely statistical issues, in contrast to other issues such as computing requirements. For an example of the quasi-repetitive behavior encountered in speech analysis, see Makhoul, 1975, Figures 1 and 4.

The point is that diversity is the hallmark of the universe of time series, and this diversity often demands a wide variety of specialized considerations often governed by the subject matter at hand, rather than a grand theory for time series. Having offered this caveat, we shall narrow our focus for the most part and deal with only a slice of the action where a particularly convenient and well-understood theory provides some useful guidelines. Namely, we shall consider discrete-parameter time series which are either homogeneous (i.e., stationary), or are reducible to such a form by fairly easy means.

Goals of Time Series Analysis. While there are many goals of time series analysis, two or three aims deserve special attention. One important goal, either for its own sake or as an adjunct to other ends, is that of constructing a prediction or forecasting function $\hat{x}_{t+\ell} = g(x_1, \ldots, x_t)$ for predicting ℓ steps (i.e., time units) ahead given the data $x_1, \ldots, x_t$. Another goal, which is basic in communications engineering, is to estimate some aspect of a signal s_t based on noisy observations $x_t = s_t + v_t$, $1 \leq t \leq n$. For example, when $s_t = \theta_1 \cos(2\pi ft+\theta_2)$, $1 \leq t \leq n$, it is desired to estimate the signal amplitude θ_1 and phase θ_2.

In geophysical, biomedical and econometric data it is often of interest to determine whether or not there exists an oscillation or cycle of some (perhaps unknown) frequency f_0, which is not immediately obvious by superficial inspection of the data. The RESX series of Figure 1b has a very obvious oscillation whose period is one year (seasonal component) and whose amplitude increases with time. Here the question is how to deal with the periodic component, not whether or not it exists. The sperm count data of Figure 2 is quite different in that it displays no obvious cycles. Yet there is an intriguing question about whether or not there is a male "period" whose biological basis is not yet understood. Although the sperm count data displays no obvious evidence in support of such a phenomenon, we shall return to this question later, armed with special methods of hunting for "cycles".

One other important use of time series analysis is to obtain reasonable inferential statements of the classical type used when the observations are supposed to be independent and identically distributed (i.i.d.). Thus one might like to construct a good 95% confidence interval for the mean of time series data, or to test the efficacy of a male contraceptive by measuring male sperm counts over periods of time prior to, and during which, the contraceptive is administered. In such situations the classical procedures, which do not take into account the dependency structure of the data over time, often give terrible results. Alternative procedures which reflect the time series structures adequately will give much more accurate results.

2. TIME SERIES MODELS

In order to do statistical inference we need to introduce formal statistical models which specify a gamut of probabilistic mechanisms which might conceivably have given rise to the observed time series $x_1, \ldots, x_n$.

Complete Descriptions. According to the most abstract version of the formalism, $x_1, \ldots, x_n$ are random variables $x_t = x_t(\omega)$, $1 \leq t \leq n$, defined on a probability space $(\Omega, \mathcal{A}, P)$ where Ω is the sample space or set of elementary outcomes, $\mathcal{A}$ is a σ-algebra of measurable subsets A of Ω and P is a probability measure which assigns a probability P(A) to each $A \in \mathcal{A}$. It is sometimes convenient to think of ω as a parameter which indexes sample paths, i.e., for each ω we get a realization or sample path $(x_1(\omega), x_2(\omega), \ldots, x_n(\omega))$ of the time series x_t. Thus the collection of elementary outcomes, i.e. the outcome space Ω, consists of all possible sample paths. At other times it is helpful to view ω as variable with t fixed so that $x_{10}(\omega)$ describes all the possible values that x_{10} can have as ω varies over all possible elementary outcomes (i.e., sample paths) in Ω.

For a variety of reasons we often regard our observations $x_1, \ldots, x_n$ as "sections" of realizations which are intrinsically singly infinite, $x_1, x_2, \ldots,$ or doubly infinite $\ldots, x_{-2}, x_{-1}, x_0, x_1, x_2, \ldots$. The time series $\{x_t\ t=1,2,\ldots\}$ or $\{x_t, t=0,\pm 1,\ldots\}$ will be denoted x_t for short, and we note that the time series x_t defined on a probability space $(\Omega, \mathcal{A}, P)$ is also called a (discrete time) random process or stochastic process.

From the probability measure we can get a less abstract but nonetheless complete description of the time series x_t by considering sets $A \in \mathcal{A}$ of the form $A = \{\omega: x_{t_1} \leq x_1, \ldots, x_{t_k} \leq x_k\}$. The joint distribution functions for the random vectors $(x_{t_1}, \ldots, x_{t_k})$ are just

$$F(x_1,\ldots,x_k;t_1,\ldots,t_k) = P(A) = P(x_{t_1} \leq x_1,\ldots,x_{t_k} \leq x_k). \tag{2.1}$$

The probability structure of the time series x_t is completely specified by specifying the above joint distribution functions for all possible collections of integer times $t_1 < \ldots < t_k$ for $k = 1,2, \ldots$. Assuming the required partial derivatives exist, we can alternatively work with the joint probability densities

$$f(x_1,\ldots,x_k;t_1,\ldots,t_k) = \frac{\partial^k}{\partial x_1 \ldots \partial x_k} F(x_1,\ldots,x_k;t_1,\ldots,t_k). \tag{2.2}$$

Stationarity. A time series x_t is said to be stationary provided all the joint distributions (2.1) are invariant under time shifts, i.e. provided

$$F(x_1,\ldots,x_k;t_1-m,\ldots,t_k-m) = F(x_1,\ldots,x_n;t_1,\ldots,t_k)$$

for all integer times $t_1 < \ldots < t_k$, $k = 1,2,\ldots$, and for all $m = \pm 1, \pm 2, \ldots$.

Gaussian Time Series. A Gaussian time series is one for which all the joint distributions (2.1) are multivariate normal. Assuming densities exist (as is the case provided linear relations with probability one do not exist among $x_1, \ldots, x_k$), this means that

$$f(x_1,\ldots,x_k;t_1,\ldots,t_k) = N(\underline{x};\ \underline{\mu},\ C) \tag{2.3}$$

with $\underline{x}^T = (x_1, \ldots, x_k)$, $\underline{\mu}^T = (Ex_{t_1}, \ldots, Ex_{t_k})$, $C_{ij} = \mathrm{cov}[x_{t_i}, x_{t_j}]$ and where

$$N(\underline{x};\ \underline{\mu},\ C) = \frac{1}{(2\pi)^{k/2}(\det C)^{1/2}} \exp(-\frac{1}{2}(\underline{x}-\underline{\mu})^T C^{-1}(\underline{x}-\underline{\mu})) \tag{2.4}$$

is the multivariate normal density function with mean vector $\underline{\mu}$ and covariance matrix C.[†]

[†] Ex_t denotes the expectation of the random variable x_t, $Ex_t = \int x\, f(x;t)dx$ and $\mathrm{cov}[x_t,x_u]$ denotes the covariance between x_t and x_u, $\mathrm{cov}[x_t,x_u] = E\{(x_t-Ex_t)(x_u-Ex_u)\} = Ex_t x_u - Ex_t Ex_u$ with $Ex_t x_u = \iint x_1 x_2 f(x_1,x_2;t,u)\, dx_1 dx_2$. In terms of the probability measure P, $Ex_t = \int x_t(\omega)dP(\omega)$ and $Ex_t x_u = \int x_t(\omega)x_u(\omega)dP(\omega)$.

For a stationary Gaussian series $Ex_t \equiv Ex_1 = \mu$ for all t, and $C_{ij} = \text{cov}[x_{t_i}, x_{t_j}] = \text{cov}[x_0, x_{t_j - t_i}] = C(t_i - t_j)$ for all i, j. Thus a stationary Gaussian time series is completely specified by its <u>mean value</u> μ and its <u>covariance function</u> $C(\ell) = E\{(x_0-\mu)(x_\ell-\mu)\}$, $\ell = 0, \pm 1, \pm 2, \ldots$.

Since the covariance function is symmetric, i.e., $C(\ell) = C(-\ell)$ for all $\ell = 0, \pm 1, \pm 2, \ldots$, it suffices to specify $C(\ell)$ for $\ell = 0, 1, \ldots$. It may be noted that $C(0) = \text{cov}[x_0, x_0] = \text{VAR } x_0 = \sigma_x^2$ is the <u>variance</u> of the series.

<u>Second Order Theory</u>. We seldom make use of the complete description of time series except in special cases such as the Gaussian one where a complete description may be given in relatively simple terms. Instead we opt for partial descriptions, and the most commonly used partial description is that consisting of the mean and covariance functions

$$\mu_t = Ex_t \tag{2.5}$$

and

$$C_{tu} = \text{cov}[x_t, x_u]. \tag{2.6}$$

When Ex_t is independent of t and C_{tu} depends only on t-u,

$$Ex_t \equiv \mu \tag{2.7}$$

$$C_{tu} = C(t-u) \tag{2.8}$$

the time series x_t is called <u>wide-sense</u> (or <u>weakly</u>) <u>stationary</u>. Clearly stationity implies wide-sense stationarity, but the converse is not true.

Let $\tilde{x}_t = x_t - Ex_t$ denote <u>centered</u> versions of the random variables x_t, t = 0, ±1, ... Then if

$$E\tilde{x}_t^2 = \int \tilde{x}_t^2(\omega)\, dP(\omega) < \infty \tag{2.9}$$

the Cauchy-Schwarz inequality yields

$$E\tilde{x}_t\tilde{x}_u = \mathrm{cov}[x_t, x_u] = \int \tilde{x}_t(\omega)\tilde{x}_u(\omega)\,dP(\omega) < \infty \tag{2.10}$$

and the random variables $\tilde{x}_t$ are elements of the Hilbert space $L^2(P)$ of random variables x which are square-integrable with respect to P; the inner product is $E\tilde{x}_t\tilde{x}_u$. This provides an elegant setting for solving prediction and other estimation problems when the second-order specification is known.

For example, one can find the solution of the minimum mean-squared-error memory—M linear predictor problem

$$\min_{a_1,\ldots,a_M} E(\tilde{x}_t - \sum_{\ell=1}^{M} a_\ell \tilde{x}_{t-\ell})^2 \tag{2.11}$$

by simply using the Hilbert Space Projection Theorem (cf. Luenberger, 1969). According to this theorem the minimizing coefficients $\bar{a}_1, \ldots, \bar{a}_M$ must be chosen such that the residual $\tilde{x}_t - \sum_{\ell=1}^{M} \bar{a}_\ell x_{t-\ell}$ is orthogonal to the subspace $\mathcal{S} = \mathcal{S}(\tilde{x}_{t-1}, \ldots, \tilde{x}_{t-M})$ spanned by $\tilde{x}_{t-1}, \ldots, \tilde{x}_{t-M}$. This gives in particular that

$$E\tilde{x}_{t-m}(\tilde{x}_t - \sum_{\ell=1}^{M} \bar{a}_\ell x_{t-\ell}) = 0, \quad m = 1,2,\ldots,M, \tag{2.12}$$

or equivalently

$$\sum_{\ell=1}^{M} C_{t-m,t-\ell}\,\bar{a}_\ell = C_{t-m,t}, \quad m = 1,2,\ldots,M, \tag{2.13}$$

which may be solved for $\bar{a}^T(M) = (\bar{a}_1, \ldots, \bar{a}_M)$.

In the case of a wide-sense-stationary series, the above equations may be written as

$$\bar{a}(M) = g \tag{2.14}$$

where C is an M×M matrix with elements $C_{ij} = C(i-j)$ and $g^T = (C(1), \ldots, C(M))$. The matrix C is a Toeplitz matrix, and this special feature of the linear equations (2.14) allows for useful computational strategies when solving for the prediction vectors $\bar{a}(M+1)$ after having already solved for $\bar{a}(M)$ (Durbin, 1960).

It should be noted that one seldom, if ever, knows the covariance structure as well as the mean values ahead of time. They need to be estimated from the data at hand as described in the next section - and so one obviously cannot use the above theory directly.

Spectral Theory and Nonparametric Models. There is an important spectral theory for wide-sense stationary time series which has as its basic ingredient the following notions.

Any real-valued function of time h(t) which is square integrable on $(-\infty,\infty)$ admits a Fourier Integral representation

$$h(t) = \int_{-\infty}^{\infty} H(f)\, e^{i2\pi tf}\, df \tag{2.15}$$

where

$$H(f) = \int_{-\infty}^{\infty} h(t)\, e^{-i2\pi ft}\, dt \tag{2.15'}$$

is called the Fourier Transform of h(t). Now let $h_m = h(m)$ be the values obtained by sampling h(t) at integer arguments $t = m = 0, \pm 1, \pm 2, \ldots$. A straightforward calculation reveals that the sequence of samples $\{h_m\}$ has the spectral representation

$$h_m = \int_{-1/2}^{1/2} \tilde{H}(f)\, e^{i2\pi mf}\, df \tag{2.16}$$

where

$$\tilde{H}(f) = \sum_{\ell=-\infty}^{\infty} H(f+\ell) \tag{2.17}$$

is periodic, with period unity. Thus $\tilde{H}(f)$ has the Fourier Series representation

$$\tilde{H}(f) = \sum_{m=-\infty}^{\infty} h_m e^{-i2\pi fm}. \qquad (2.18)$$

Since the h_m are real, $\tilde{H}(f) = \tilde{H}^*(-f)$ and we have the following finite-sum approximation for h_m:

$$h_m \cong \frac{1}{2N} \sum_{j=-N}^{N} \tilde{H}(f_j)\, e^{i2\pi f_j m}, \qquad f_j = j/2N \qquad (2.19)$$

$$\cong A_0 + \sum_{j=1}^{N} A_j \cos(2\pi f_j m + \Phi_j). \qquad (2.20)$$

Thus the <u>deterministic</u> discrete-time sequence $\{h_m\}$ is approximately a finite sum of cosines of <u>frequencies</u> $f_j = j/2N$, <u>amplitudes</u> A_j and <u>phases</u> Φ_j $(\Phi_0 = 0)$.

Now it turns out that any wide-sense stationary discrete time series $\{x_m\}$ having <u>zero mean</u> and <u>finite variance</u> admits a similar representation. Thus

$$x_m \cong \sum_{j=-N}^{N} \Delta Z(f_j)\, e^{i2\pi f_j m}, \qquad f_j = j/2N \qquad (2.21)$$

$$= \sum_{j=1}^{N} B_j \cos(2\pi f_j m + \Phi_j) \qquad (2.22)$$

where the $\Delta Z(f_j) = Z(f_j + \frac{1}{2N}) - Z(f_j - \frac{1}{2N})$ and B_j are <u>random</u> coefficients and the Φ_j are <u>random</u> phases. Furthermore, the B_j are uncorrelated, i.e. $\mathrm{cov}[B_i, B_j] = 0$, $i \neq j$. The exact limiting version of these approximations is the <u>spectral representation</u> (Lamperti, 1977)

$$x_m = \int_{-1/2}^{1/2} e^{i2\pi mf}\, dZ(f) \qquad (2.23)$$

where the <u>spectral process</u> $Z(f)$, $-1/2 < f \leq 1/2$, satisfies $Z(-1/2) = 0$ a.s., and is a random process with <u>orthogonal increments</u>:

$$E[Z(f_4)-Z(f_3)][Z(f_2)-Z(f_1)]^* = 0, \qquad f_1 < f_2 < f_3 < f_4 . \tag{2.24}$$

The covariance sequence $C(\ell)$, $\ell = 0, \pm 1, \pm 2, \ldots$, is readily obtained from (2.23) using the orthogonal increments property of $Z(f)$:

$$C(\ell) = \operatorname{cov}[x_m, x_{m+\ell}] = \int_{-1/2}^{1/2} e^{i2\pi\ell f}\, dF(f) \tag{2.25}$$

where the <u>spectral distribution function</u> F is a monotonically increasing function, with $F(-1/2) = 0$ and $F(1/2) = C(\ell) = \operatorname{Var} x_m$. F is related to the spectral process $Z(f)$ by

$$F(f) = E|Z(f)|^2 \tag{2.26}$$

and

$$dF(f) = E|dZ(f)|^2. \tag{2.27}$$

Assuming that F does not have a singular component, F may be decomposed into its discrete and absolutely continuous parts:

$$F = F_d + F_{ac}. \tag{2.28}$$

F_d is a right-continuous step function and the absolutely continuous part F_{ac} has a derivative $S(f) = (d/df)F(f)$ which is called the <u>spectral density</u>. When $F_d \equiv 0$ the spectral density $S(f)$ can be obtained by inverting (2.25):

$$S(f) = \sum_{\ell=-\infty}^{\infty} C(\ell) e^{-i2\pi f\ell}. \tag{2.29}$$

The following are two important examples. If $x_m = B\cos(2\pi f_0 m + \Phi)$ with B

and Φ independent, $EB^2 < \infty$, and Φ uniformly distributed on $[-\pi,\pi]$, then x_m is a wide-sense stationary process for which the approximation (2.22) is exact (with $B_0 \equiv 0$, $N = 1$, $B_1 = B$). In this particularly simple case $F = F_d$ is discrete with just two jumps of height $EB^2/4$ at $\pm f_0$. The second example is where $F = F_{ac}$ is absolutely continuous with rational spectral density

$$S(f) = \sigma^2 \frac{|\sum_{j=0}^{q} \theta_j e^{i2\pi fj}|^2}{|\sum_{j=0}^{p} \phi_j e^{i2\pi fj}|^2}, \quad \theta_0 = \phi_0 = 1. \tag{2.30}$$

We shall return to this important case shortly.

It may be noted that since the B_j in (2.22) are related to $Z(f)$ by $B_j = 2|\Delta Z(f_j)|$, we have the following important heuristic relationship between the approximating coefficients B_j and the spectral density:

$$EB_j^2 \propto S(f_j). \tag{2.31}$$

Thus when one estimates a spectral density, for example as described in the next section, one is in effect estimating the mean-squared values of the random amplitudes B_j. The random phases are lost in the spectral distribution function (and density) because of the squared magnitude in (2.27).

Wold's Decomposition. The specification of a time series by its spectral distribution function is a nonparametric specification in the sense that the spectral distribution function is not in general defined in terms of a finite number of parameters. Corresponding to the nonparametric frequency domain modeling of a time series in terms of its spectral distribution function, there is a time domain nonparametric model due to H. Wold (1954).

Wold's result is that any zero-mean wide-sense stationary time series x_m may be decomposed in the form

$$x_m = x_m^d + x_m^{ac} \tag{2.32}$$

with x_m^d and x_m^{ac} being mutually orthogonal processes; furthermore,

$$x_m^{ac} = \sum_{\ell=0}^{\infty} h_\ell \varepsilon_{m-\ell} \tag{2.33}$$

where $\{e_j\}$ is a sequence of uncorrelated random variables, called innovations, with common means of zero and common variances σ_ε^2. The process x_m^d corresponds to the singular part F_d of the spectral distribution function and x_m^{ac} corresponds to the absolutely continuous part F_{ac}. The process x_m^d is perfectly predictable, as is clear in the special case $x_m = B\cos(2\pi f_0 m + \Phi)$ given above. For the most part we assume that x_m contains no perfectly predictable part, so that $x_m = \sum_{\ell=0}^{\infty} h_\ell \varepsilon_{m-\ell}$. Since this time domain model is specified in terms of an infinite number of parameters σ_ε^2, h_0, h_1, ..., it is regarded as a nonparametric model.

Parametric Models. Probability models which are completely specified in terms of a finite number of parameters are called parametric models. Thus if $\varepsilon_m \sim N(0,\sigma_\varepsilon^2)$ the finite moving-average model

$$x_m = \mu + \varepsilon_m + \theta_1\varepsilon_{m-1} + \ldots + \theta_q\varepsilon_{m-q} \tag{2.34}$$

is a parametric model which is completely specified by the parameters $\theta_1, \theta_2, \ldots, \theta_q$, σ_ε^2 and μ, the last being the mean value of x_m. This model is obviously a special case of the infinite moving average (2.33).

A most important type of parametric time series model is the so-called autoregressive-moving-average (ARMA) model of order (p,q)

$$x_m - \phi_1 x_{m-1} \cdots - \phi_p x_{m-p} = \gamma + \varepsilon_m + \theta_1\varepsilon_{m-1} + \ldots + \theta_q\varepsilon_{m-q} \tag{2.35}$$

where now $Ex_m = \mu = \gamma/(1-\sum_1^p \phi_i)$. This model will be strictly speaking parametric only if the distribution of ε_m is specified by a finite number of parameters. The usual assumption is that $\varepsilon_m \sim N(0,\sigma_\varepsilon^2)$.

In order for x_m to be a wide-sense stationary process it is necessary and sufficient that all the roots of the characteristic polynomial $d(z) = 1 - \phi_1 z^{-1} \ldots - \phi_p z^{-p}$ be inside the unit circle (Anderson, 1971). In addition, it is usually assumed that all the roots of the polynomial $n(z) = 1 + \theta_1 z^{-1} \ldots + \theta_q z^{-q}$ are inside the unit circle. The first of these two conditions insures that the autoregressive part of (2.35) can be "inverted" so that x_m has an infinite moving-average representation of the form (2.33). The second condition insures that the moving-average part of (2.35) can be "inverted" so that x_m also has an <u>infinite autoregressive</u> representation

$$x_m - g_1 x_{m-1} - g_2 x_{m-2} \ldots = \delta + \varepsilon_m . \tag{2.36}$$

The latter property is quite useful in connection with parameter estimation and forecasting.

It should be noted that the first-order moving-average series

$$x_m = \varepsilon_m + \theta \varepsilon_{m-1} \tag{2.37}$$

has the infinite-order autoregressive representation

$$x_m - \theta x_{m-1} + \theta^2 x_{m-2} - \theta^3 x_{m-3} \ldots = \varepsilon_m \tag{2.38}$$

provided $|\theta| < 1$. However, the moving-average form is preferred on grounds of parsimony. Similarly, the autoregression $x_m - \phi x_{m-1} = \varepsilon_m$ has the infinite moving-average form $x_m = \varepsilon_m + \phi \varepsilon_{m-1} + \phi^2 \varepsilon_{m-2} \ldots$, but we prefer the autoregression form.

It may be shown that when $\gamma = 0$ the ARMA process (2.35) has the spectral density function given by (2.30). Since any spectral density $S_0(f)$ which is continuous on $[-1/2, 1/2]$ can be approximated uniformly well by polynomials in $\cos 2\pi f$, uniformly good rational approximation of such $S_0(f)$ by the ARMA spectral density (2.30) is clearly possible. Use of the rational form, rather than just using the numerator in (2.30), generally allows for a good approximation with fewer coefficients.

3. MODEL ESTIMATION

Corresponding to the parametric and nonparametric models for time series, there are parametric and nonparametric methods of estimating the models from the data.

Parametric Estimation. If the data $x_1, \ldots, x_n$ were truly generated according to a Gaussian ARMA(p,q) model, then we would naturally want to estimate the parameters $\underline{\phi} = (\phi_1, \ldots, \phi_p)$, $\underline{\theta} = (\theta_1, \ldots, \theta_q)$, σ_ε^2 and γ by maximizing the Gaussian likelihood*

$$L(\underline{\phi},\underline{\theta},\sigma_\varepsilon^2,\gamma) = f(x_1,x_2,\ldots,x_n;\underline{\phi},\underline{\theta},\sigma_\varepsilon^2,\gamma) \tag{3.1}$$

where $f(x_1, \ldots, x_n; \underline{\phi}, \underline{\theta}, \sigma_\varepsilon^2, \gamma)$ is the multivariate normal density for the observations $\underline{x} = (x_1, \ldots, x_n)$. It may be shown that the log-likelihood $\mathcal{L}(\underline{\phi}, \underline{\theta}, \sigma_\varepsilon^2, \gamma) = \log f(\underline{x}; \underline{\phi}, \underline{\theta}, \sigma_\varepsilon^2, \gamma)$ is approximated by

$$\tilde{\mathcal{L}}(\underline{\phi},\underline{\theta},\sigma_\varepsilon^2,\gamma) = -n \log \sigma_\varepsilon - \frac{S(\underline{\phi},\underline{\theta},\gamma)}{2\sigma_\varepsilon^2} \tag{3.2}$$

up to an additive constant, where

$$S(\underline{\phi}',\underline{\theta}',\gamma') = \sum_{m=1}^{n} \hat{\varepsilon}_m(\underline{\phi}',\underline{\theta}',\gamma')^2 \tag{3.3}$$

is the sum-of-squares function with the $\hat{\varepsilon}_m = \hat{\varepsilon}_m(\underline{\phi}',\underline{\theta}',\gamma')$ being the estimates of the ε_m based on parameter values $\underline{\phi}'$, $\underline{\theta}'$, γ', and the data $x_1, \ldots, x_n$. These

*Maximum-likelihood estimates are consistent and asymptotically efficient under regularity conditions.

estimates are computed recursively using (2.35):

$$\hat{\varepsilon}_m = x_m - \phi_1' x_{m-1} \cdots - \phi_p' x_{m-p} - \gamma' - \theta_1' \hat{\varepsilon}_{m-1} \cdots - \theta_q' \hat{\varepsilon}_{m-q} \tag{3.4}$$

with appropriate initial conditions.

In view of the approximation (3.2), approximate maximum-likelihood estimates $\hat{\underline{\phi}}$, $\hat{\underline{\theta}}$, $\hat{\gamma}$ are obtained by minimizing the sum-of-squares function $S(\underline{\phi}', \underline{\theta}', \gamma')$ with respect to its arguments. Due to the presence of moving-average terms the $\hat{\varepsilon}_m$ are nonlinear functions of the parameters, which means that the desired minimization results in a nonlinear least-squares problem. Details concerning the nonlinear structure of $S(\underline{\phi}', \underline{\theta}', \gamma')$ and numerical methods for minimizing this function now abound in the literature (see Nelson, 1973 and Box and Jenkins, 1976 for a good start). In fact, recently some attention has been devoted to the computation of exact maximum-likelihood estimates of ARMA model paraemters (Phadke and Kedem, 1978; Ansley, 1979).*

The special case of autoregression deserves some additional comment. First of all, the sum-of-squares function for autoregressions is

$$S(\underline{\phi}', \gamma') = \sum_{m=2}^{n} (x_m - \gamma' - \textstyle\sum_1^p x_{m-\ell} \phi_\ell')^2 \tag{3.5}$$

where $x_0 = x_{-1} = \cdots = x_{-p+2} = 0$. Thus the minimization of $S(\underline{\phi}', \gamma')$ results in the linear "normal" equations

$$\begin{aligned} &\sum_{m=2}^{n} (x_m - \hat{\gamma} - \textstyle\sum_1^p x_{m-\ell} \hat{\phi}_\ell) = 0 \\ &\sum_{m=2}^{n} x_{m-j} (x_m - \hat{\gamma} - \textstyle\sum_1^p x_{m-\ell} \hat{\phi}_\ell) = 0, \qquad 1 \le j \le p. \end{aligned} \tag{3.6}$$

The term autoregression is natural in view of the fact that (2.35), with $\theta_1 = \theta_2 = \cdots = \theta_q = 0$, may be written in the usual linear model form, with intercept γ:

$$y_m = \underline{z}_m \underline{\beta} + \varepsilon_m, \qquad 2 \le m \le n \tag{3.7}$$

---

*Such estimates may be useful when the "true" parameter values yield roots of the polynomials $d(z)$, $n(z)$ which are close to the unit circle, thereby rendering $\mathcal{L}(\underline{\phi}, \underline{\theta}, \sigma_\varepsilon^2, \gamma)$ a poor approximation to the exact likelihood.

where $y_m = x_m$, $\underline{z}_m = (1, x_{m-1}, \ldots, x_{m-p})$, $\underline{\beta}^T = (\gamma, \phi_1, \ldots, \phi_p)$. In matrix form we have

$$\underline{y} = Z\underline{\beta} + \underline{\varepsilon} \tag{3.8}$$

where $\underline{y}$, $\underline{\varepsilon}$ are $(n-1)\times 1$ vectors, $\underline{\beta}$ is a $(p+1)\times 1$ vector and Z is a $(n-1)\times(p+1)$ matrix. Then (3.6) are recognizable as the normal equations

$$Z^T Z\hat{\underline{\phi}} = Z^T \underline{y} \tag{3.9}$$

where Z contains an upper-right triangle of zeros corresponding to setting $x_0 = x_{-1} = \ldots = x_{-p+2} = 0$.

Small but sometimes important variants of the estimate $\hat{\underline{\phi}}$ are obtained by different specifications of Z. Often the first row of Z is taken to be $\underline{z}_{m+1}$ so that a full set of predictor variables is available; Z is then an $(n-p)\times(p+1)$ matrix.

Neither this definition of Z nor the one given by (3.7) - (3.8) are guaranteed to yield estimated coefficients which correspond to a stationary autoregression, as would be quite desirable in some applications. However, it is well known how to set up somewhat different linear estimating equations which do yield estimates $\hat{\phi}_1, \ldots, \hat{\phi}_p$ which correspond to a stationary process, and which also allow for efficient computation of the coefficient estimates for a sequence of models of orders $p = 1, 2, \ldots, p_{max}$ (see for example Box and Jenkins, 1976; Robinson, 1967). However, these estimates are not without their own special problems (Martin, 1980a).

Order Determination for Parametric Models. One of the difficulties encountered when fitting ARMA models to data is that one does not know the orders p and q in advance (indeed, the use of an ARMA model is at best an adequately good approximation to reality). The order determination problem is often called the identification problem. Box and Jenkins (1976) and Nelson (1973) describe one approach to determining the orders p and q.

More recently Akaike (1974) has proposed attractive methods for order determination which are based on minimizing the quantity $-2 \log f(\underline{x};\hat{\underline{\phi}},\hat{\underline{\theta}},\hat{\sigma}_\varepsilon^2,\hat{\gamma}) + c(p,q)$ where $\hat{\underline{\phi}}$, $\hat{\underline{\theta}}$, $\hat{\sigma}_\varepsilon^2$, $\hat{\gamma}$ are the maximum-likelihood estimates (or approximations thereof in practice) for an ARMA model of order (p,q), and $c(p,q) = 2(p+q+2)$. This criterion accounts for bias as well as variability, and the correction term in effect reflects a penalty for overfitting at a given fixed sample size. A recent unpublished study by Geweke and Meese (1979) provides an up-to-date account of order selection proposals due to Akaike and others as well. For the special case of autoregression, one might wish to read the papers by Jones (1974) and Shibata (1976).

The Autoregressive Spectral Density Estimate. A natural parametric estimate of the spectral density may be obtained by substituting the estimates of autoregressive parameters in place of σ^2, $\phi_1, \ldots, \phi_p$ in (2.30) with $\theta_1 = \ldots = \theta_q = 0$:

$$\hat{S}(f) = \frac{\hat{\sigma}_\varepsilon^2}{\left|\sum_{j=0}^{p} \hat{\phi}_j e^{i2\pi fj}\right|^2}, \qquad \phi_0 = 1. \tag{3.10}$$

A good approximating order $\hat{p}$ may be obtained using a good order selction rule such as Akaike's criterion described above or Parzen's CAT criterion (1974). Some pertinent references on autoregressive spectral density estimates are Akaike (1969), Parzen (1974) and Jones (1974).

Nonparametric Estimation. Although the covariance function $C(\ell)$ vanishes for $|\ell| > q$ if x_m is a qth-order moving average, this is not generally the case, and even if it is the case one generally will not know it in advance. Thus, since we are usually led to consider the full infinite sequence $C(\ell)$, $\ell = 0, 1, 2, \ldots$, the second-order specification of a wide-sense stationary time series in terms of its covariance sequence (along with its mean value) is regarded as nonparametric.

One natural estimate of $C(\ell)$ is

$$\hat{C}_0(\ell) = \frac{1}{n-\ell} \sum_{m=1}^{n-\ell} (x_m - \bar{x})(x_{m+\ell} - \bar{x}), \qquad 0 \leq \ell \leq n-1 \tag{3.11}$$

where $\bar{x} = (1/n) \sum_1^n x_i$. However, the estimate

$$\hat{C}(\ell) = \frac{1}{n} \sum_{m=1}^{n-\ell} (x_m - \bar{x})(x_{m+\ell} - \bar{x}), \qquad 0 \leq \ell \leq n-1 \tag{3.12}$$

is generally preferred because the quadratic form $\sum_{i,j=1}^n b_i b_j \hat{C}(i-j)$ is positive definite, and the matrix $\hat{C}$ is Toeplitz. As a result, use of the $\hat{C}(\ell)$ in place of the $C(\ell)$ in (2.14) yields predictor-coefficient estimates $\hat{a}(M)$ which correspond to a stationary Mth-order autoregression (Robinson, 1969). These equations are the alterations of (3.9) which were alluded to earlier.

The most frequently used nonparametric method of estimating the spectrum today is based on smoothing the periodogram. Given the observed time series x_k, $1 \leq k \leq n$, the basic steps in constructing the estimate are as follows. A data window or taper d_k, $1 \leq k \leq n$, which slopes gently towards zero at both ends, is used to form the modified data $\bar{x}_k = d_k x_k$. The purpose of this tapering is to reduce "leakage" (see e.g. Bloomfield, 1976) associated with the finiteness of any actual set of data. Then the discrete Fourier transform of $\bar{x}_k$

$$X(f_\ell) = \sum_{k=1}^{n} \bar{x}_k \exp(-\frac{i2\pi k\ell}{n}), \qquad f_\ell = \ell/n, \quad \ell = 0,1,\ldots,[\tfrac{n}{2}] \tag{3.13}$$

is computed using the fast Fourier transformation, and the periodogram

$$\hat{S}(f_\ell) = \frac{1}{n} |X(f_\ell)|^2 \tag{3.14}$$

is formed. Since the periodogram is not a consistent estimate, it is smoothed to get a respectable estimate of the power spectrum:

$$\bar{S}(f_\ell) = \sum_{j=-L}^{L} w_j \hat{S}(f_{\ell+j}) \tag{3.15}$$

$$w_j = w_{-j}, \quad \sum_{j=-L}^{L} w_j = 1.$$

It is also possible to reduce variability by averaging periodograms computed from different (often overlapping) time segments (see Welch, 1967).

A good spectrum analysis by the above method usually requires computing several estimates based on different amounts of smoothing in order to exploit the tradeoffs between bias and variance. The leakage problem can be alleviated by either applying one of the better tapering functions such as the cosine window, the Parzen window or the prolate spheroidal window (see Thomson, 1977) or through prewhitening (Kleiner, Martin and Thomson, 1979).

Further details concerning the above "direct" method (by way of contrast with the pre-Fast Fourier Transform approach based on using covariance estimates $\hat{C}(\ell)$, $\ell = 0, 1, \ldots, L$ in (2.29)) may be found in the introductory text by Bloomfield (1976).

Reduction to Stationarity. Many time series encountered in practice hardly resemble a stationary series. For example, the trendy behavior exhibited by the series in Figure 1a is a fairly common occurrence, as is the trendy seasonal oscillation displayed in Figure 1b. Thus something needs to be done to reduce such series to stationarity, or to remove such stationarity in order to apply the model estimation techniques just discussed.

The most simple and commonly used device for doing this is to take differences. Thus one might form the first differences

$$\nabla x_m = x_m - x_{m-1}, \quad 2 \le m \le n \tag{3.16}$$

in order to remove trends such as that of Figure 1a. The results are displayed in Figure 3. Seasonal data like RESX in Figure 1b requires differencing of lag-12:

$$\nabla_{12} x_m = x_m - x_{m-12}, \qquad 13 \leq m \leq n \tag{3.17}$$

to remove the seasonal component. The result is shown in Figure 4 . Similarly one might remove the seasonal components, such as quarterly movements of economic time series, by appropriate differencing. Further details may be found in Nelson (1973) and Box and Jenkins (1976). For the removal of stationarity by fitting polynomials, sines and cosines, the reader is referred to Anderson (1971).

The problem of adjusting for (i.e., estimating and removing) trend and seasonal components is called the seasonal adjustment problem, and it is of considerable importance to diverse organizations such as the Bureau of Labor Statistics, etc. For a modern approach to seasonal adjustment procedures which are robust (cf. Section 5), the reader may refer to Cleveland, Dunn and Terpenning (1978a, 1978b).

4. DATA ANALYTIC TECHNIQUES AND EXPLORATORY TOOLS

One hallmark of present-day statistical activities is the increased attention being devoted to data analysis, and to the development of new methodologies for examining the structure of data in an informative exploratory manner (see, for example, Tukey, 1977 and Gnanadesikan, 1977). Well-chosen graphical displays serve well for the latter purpose. Furthermore, formal statistical models play a diminished role in broadly construed exploratory work, unlike as in confirmatory data analysis (Tukey, 1977) where formal models provide indispensible guidance for constructing statements of precision (e.g., confidence intervals). We shall illustrate some simple exploratory tools by example in the time-series context.

Box Plots. Box plots, introduced by J. W. Tukey (1977), are displayed in Figure 5 for sperm-count time-series data from six different individuals (cases). The upper and lower ends of the boxes correspond (roughly) to the upper and lower quartiles

UQ and LQ of the data, and the horizontal line in the middle of the box is the median of the data. The ends of the vertical bars are located at data points which are farthest from the median of all those data points which are at a distance less than 1.5 × (UQ-LQ) from the median. The remaining "outside" points are also displayed. Finally, the width of the box is proportional to the square root of the sample size used to construct the box.

Obviously, the box plot provides a univariate distributional view of a time-series data set which corresponds to projecting the time plot onto the vertical axis. This clearly entails the loss of potentially useful second- (and higher-) order information contained in the original time-ordered data. On the other hand, the display of box plots still allows one to make easy comparison of important univariate distributional information (i.e., the center, width, skewness, and extreme points of the data). The following comments pursue the skewness issue a bit further.

Data Transformations. Many kinds of data, including time-series data, fail to have even univariate normal distributions. However, this failure often manifests itself in forms of skewness which allow for easy transformation to near-normality or at least near-symmetry. Inspection of the box plots in Figure 5 reveals a tendency in the data of skewness toward large values. Application of a logarithmic transformation to the original sperm count series results in the box plots of Figure 6. Clearly the logarithmic transformation has gone too far; the data now appears skewed toward small values. The square root transformation is more gentle, and its use results in the agreeably symmetric box plots of Figure 7.

Tukey (1977) gives a nice data-oriented view of the hierarchy of simple transformations, and Box and Cox (1964) discuss an important class of transformations and the problem of selecting an appropriate transformation for the data at hand.

Quantile-Quantile Plots. Quantile-quantile or Q-Q plots (Gnanadesikan, 1977) are variants of probability plots which reveal more detail than box plots about deviations of the data from an assumed distributional model. A normal Q-Q plot is a scatter plot of the pairs $(x(p_i), x_{(i)})$, $1 \leq i \leq n$ where $x_{(1)} \leq x_{(2)} \leq \cdots \leq x_{(n)}$ are the order statistics and $x(p_i) = \Phi^{-1}(p_i)$ is the p_i-quantile of the standard normal distribution Φ, with $p_i = (i - 1/2)/n$, $1 \leq i \leq n$. For normal data the points in the scatter plot will tend to be close to a straight line.

Normal Q-Q plots for the sperm-count data sets were much clearer indicators of non-normality of both the original series and the logarithmically transformed series than were the box plots. Figure 8 displays the normal Q-Q plot for the square-root transformed data. The nearly linear configuration of the plot is indicative of a reasonably normal data set (from a univariate point of view).

Correlation Function Plots. The estimated correlation function (also called the autocorrelation function or correlogram) is $\hat{\rho}(\ell) = \hat{C}(\ell)/\hat{C}(0)$, where $\hat{C}(\ell)$ is the estimated covariance function. Clearly $\hat{\rho}(0) = 1$ and the Cauchy-Schwartz inequality shows that $|\hat{\rho}(\ell)| \leq 1$ for all $\ell \leq n$. It is a standard practice to plot $\hat{\rho}(\ell)$ for time-series data sets, and to superimpose 90 or 95% confidence limits for the null hypothesis that the true correlation function is zero for $\ell > 0$ (Nelson, 1973; Box and Jenkins, 1976). Although these plots are often useful in indicating serial dependence in the data, they have not been quite as useful in practice as one might have hoped.

Residuals Plots. As in ordinary regression analysis, a variety of plots based on estimated ARMA-model residuals $\hat{\varepsilon}_m = \hat{\varepsilon}_m(\hat{\underline{\phi}}, \hat{\underline{\theta}}, \hat{\gamma})$ can be extremely useful in assessing the adequacy of the model. If the model fits well, then the residuals $\hat{\varepsilon}_m$ should be nearly uncorrelated, and this would be revealed by a plot of the correlation function estimate $\hat{\rho}_\varepsilon(\ell)$ based on these residuals. Lack of a good fit is often revealed by large values of $\hat{\rho}_\varepsilon(\ell)$ for $\ell \neq 0$. Some pertinent examples may be found in the Nelson and Box and Jenkins references.

As a supplement, or possibly even alternative, to plotting $\hat{\rho}_{\varepsilon}(\ell)$, I would suggest routinely making scatter plots of residuals $\hat{\varepsilon}_m$ against lag-ℓ residuals $\hat{\varepsilon}_{m-\ell}$ for $\ell = 1, \ldots, L$ for some appropriate maximum lag L. Often the single plot of $\hat{\varepsilon}_m$ against $\hat{\varepsilon}_{m-1}$ will be quite revealing. The near-uncorrelatedness of the $\hat{\varepsilon}_m$ for a well-fitting model will be reflected by a circular shape of the scatter plot. Furthermore, if there are any outliers (i.e., unusually large data points) which may cause trouble for both the ARMA-parameter estimates and $\hat{\rho}_{\varepsilon}(\ell)$, they will be clearly revealed.

By way of example we display in Figure 9 a plot of $\hat{\varepsilon}_m$ versus $\hat{\varepsilon}_{m-1}$ for the least-squares fit of a second-order autoregression to the seasonally-differenced RESX data of Figure 4. The curious deviation from circularity of the bulk of the data indicates possible lack of good fit, and the outliers in evidence are also clearly displayed in the differenced series of Figure 4. It turns out that the outliers are causing the poorness of fit, and that a second-order autoregression would be adequate if it were not for the outliers. Further details concerning this example may be found in Martin (1980a).

One could also make a check on the normality assumption for the ε_m, which underlies the use of the Gaussian maximum-likelihood estimate, by making normal Q-Q plots of the residuals $\hat{\varepsilon}_m$.

Spectral Density Estimates. Spectral density estimates are in fact most often used as exploratory tools whose main messages are conveyed by the special features of their shapes. Although normal theory confidence intervals are often reported with spectral density estimates (cf. Bloomfield, 1976, p. 195), it is relatively infrequently that an investigator proceeds very far with formal statistical inference.

A sixth-order autoregressive spectral density estimate for the square-root-transformed sperm count data of Figure 2 is displayed in Figure 10. The peaks at .17 and .38 cycles/week correspond to the rapid wiggles superimposed on the slower trendy behavior of the time plot. The peak near the origin of the spectrum corresponds to the dominant slow trend of the data. No attempt has been made to use autoregressive order-selection rules. However, three of the six cases at hand exhibited two peaks in the sixth-order autoregressive spectrum (located at about the same frequencies as in Figure 10), which is suggestive enough to warrant further careful study of these six cases plus the much larger number of cases available.

5. OUTLIERS AND ROBUST PROCEDURES

Robustness Concepts. Loosely speaking, a robust estimate is one whose performance remains quite good when the true distribution of the data deviates from the assumed distribution. Data sets for which one often makes a Gaussian assumption sometimes contain a small fraction of unusually large values or "outliers" (Figures 1a and 1b provide examples). More realistic models for such data sets are provided by distributions which are heavy-tailed relative to a nominally Gaussian distribution. Quite small deviations from the Gaussian distribution, in an appropriate metric, can give rise to heavy-tailed distributions and correspondingly potent outliers in the generated samples. Since such deviations from a Gaussian distribution can have serious adverse effects on Gaussian maximum-likelihood procedures, it is not surprising that attention has been focused primarily on robustness over families of distributions which include both the nomrinal Gaussian model and heavy-tailed distributions.

The seminal papers by Tukey (1960), Huber (1964) and Hampel (1971) provide more precise notions of robustness, namely efficiency robustness, min-max robustness and qualitative robustness, respectively.

An efficiency robust estimate is one whose (appropriately defined) efficiency is "high" at a variety of strategically chosen distributions. Huber's min-max robust location estimates minimize the maximum asymptotic variance over certain uncountably infinite families of distributions.

Hampel's qualitative robustness is a natural equicontinuity requirement. Suppose F_0, $\mathbf{F}$ are distributions for i.i.d. observations, $\{T_n\}$ a sequence of estimators indexed by sample size, and $\mathcal{L}(T_n, F_0)$, $\mathcal{L}(T_n, F)$ the distributions (laws) of T_n under F_0, F respectively. Then $\{T_n\}$ is qualitatively robust at F_0 if the sequence of maps $F \to \mathcal{L}(T_n, F)$ is equicontinuous at F_0 in an appropriate metric.

Robustness Concepts for Time Series. The robustness concepts just described have been developed and used primarily in the context of independent and identically distributed observations (possibly vector-valued as when estimating covariance matrices).

For time series parameter estimation problems, efficiency robustness and min-max robustness are concepts directly applicable.

A major problem which remains is that of providing an appropriate and workable definition of qualitative robustness in the time-series context. Some results in this direction have been provided by Kazakos and Gray (1977).

Time Series Outlier Models. Formal modeling of outliers in time series is a task made particularly challenging by the dependency features one must work with, and the fact that time series "outliers" often occur not just in isolation but in patches which have their own local correlation structure. One simple outlier model which seems to have reasonable utility is the following additive outlier (AO) model:

$$y_m = \mu + x_m + v_m \tag{5.1}$$

where μ is a location parameter, x_m is a Gaussian ARMA model and v_m are the additive outliers which are non-zero a small fraction of the time, i.e., $P(v_m = 0) = 1-\gamma$ with γ small. At this level of generality the v_m could be either isolated or patchy outliers. Further discussion of outliers models, including an alternative innovations outliers model, may be found in Martin (1979a, 1980a, 1980b).

A Robust ARMA Model-Fitting Procedure. We know from P. Huber's work (1964, 1977) that maximum-likelihood estimates for appropriately chosen heavy-tailed outlier generating distributions (which have a central Gaussian shape) are efficiency robust and qualitatively robust. The algorithmic essence of an approximate maximum-likelihood type estimate of ARMA model parameters for the additive outliers model (with i.i.d. v_m) is as follows:

1) Use a robust filtering algorithm based on preliminary estimates $\underline{\hat{\phi}}^0$, $\underline{\hat{\theta}}^0$, $\hat{\sigma}_\varepsilon^0$, $\hat{\gamma}^0$ to transform the observations y_m to filtered values $\hat{x}_m^0$.
2) Use the $\hat{x}_m$ in a Gaussian M.L.E./least-squares algorithm to obtain new estimates $\underline{\hat{\phi}}^1$, $\underline{\hat{\theta}}^1$, $\hat{\sigma}_\varepsilon^1$, $\hat{\mu}^1$.
3) Iterate this procedure by going back to step one, using the new parameter estimates in place of those used in the previous iteration, until there is little change in the estimates.

Details for the autoregressive case may be found in Martin (1979a), and in Kleiner, Martin and Thomson (1979). The robust estimation procedure for ARMA models is discussed in Martin (1980b).

For a well-tuned robust estimation algorithm of the type described above, it typically turns out that $\hat{x}_m = y_m$ most of the time, and extreme outliers are replaced by one-sided interpolates (two-sided outliner interpolation is described in Martin, 1979b).

One example of using the above algorithm is provided in Figure 11, which shows the original seasonally differenced RESX data with the filter values $\hat{x}_m$ superimposed. Indeed, $\hat{x}_m$ coincides with the original series most of the time, and the two huge outliers are replaced by one-sided interpolates. In addition, two other locally high data points are altered.

It might be argued that such obvious outliers as those in the RESX data could easily be spotted and "appropriately" altered by the careful statistician. While this indeed may sometimes be possible, it will still be necessary to specify an algorithm for modifying bad data points when using the computer. It is not too surprising that an approximate non-Gaussian maximum-likelihood estimate involves the use of a "data-cleaning"/"outlier-interpolator" algorithm whose mode of operation is intuitively appealing and rougly does what the "by-the-eyeball" statistician might do in practice anyway, at least for obvious outliers.

A much more subtle example is given in Figure 4a of Kleiner, Martin and Thomson (1979), where the outliers are not visible in a usual plot of the data. However, their impact on wide-dynamic-range spectral analysis via smoothed periodograms can be quite significant, as Figure 4c of the cited reference shows. That figure also displays a considerably improved spectral density estimate which was obtained using a robust filtering technique.

It may be noted that for the case of autoregression one may obtain robust estimates by using natural analogues (Denby and Martin, 1979; Martin, 1979a; Martin, 1980a) of bounded influence regression estimates (see Welch, 1979, for an overview of such regression estimates, including references to the important earlier work of F.R. Hampel and C.L. Mallows). The asymptotic behavior of bounded-influence autoregression estimates is currently better understood than that of the approximate M.L.E.s described above.

6. CONCLUDING COMMENT

In addition to perusing the more or less standard literature on time series, I would suggest that those who become seriously interested in the subject should read J.W. Tukey's (1980) unique and stimulating paper "Can we predict where time series should go next?"

ACKNOWLEDGEMENTS

The figures presented in this paper were made using the S-system for interactive data analysis developed by Becker and Chambers (1978); see also Chambers (1980).

This paper was prepared with support from NSF Grant SOC 78-09474.

REFERENCES

Akaike, H. (1969), "Power spectrum estimation through autoregressive model fitting," Ann. Instit. Statist. Math., 21, 243-247.

Akaike, H. (1974), "A new look at the statistical identification", IEEE Trans. on Auto. Control, 19, 723-730.

Anderson, T.W. (1971), The Statistical Analysis of Time Series, Wiley, New York.

Ansley, C.F. (1979), "An algorithm for the exact likelihood of mixed autoregressive-moving average process", Biometrika, 66, 59-65.

Bartlett, M.S. (1978), An Introduction to Stochastic Processes, Cambridge Univ. Press, New York.

Becker, R.A., and Chambers, J.M. (1978), "Design and implementation of the S-system for interactive data analysis", Proceedings of Compsac 78, Computer Software and Application Conf., IEEE.

Bloomfield, P. (1976), Fourier Analysis of Time Series, Wiley, New York.

Box, G.E.P. and Cox, D.R. (1964), "An analysis of transformations", Jour. Roy. Stat. Soc., B, 26, 211-252.

Box, G.E.P. and Jenkins, G.M. (1976), Time Series Analysis: Forecasting and Control, Holden-Day, San Francisco.

Chambers, J.M. (1980), "Statistical computing: past history and future trends", American Statistician (to appear).

Cleveland, W.S., Dunn, D.M. and Terpenning, I.J. (1978a), "A resistant seasonal adjustment procedure with graphical methods for interpretation and diagnosis", in Seasonal Analysis of Economic Time Series, edited by Arnold Zellner, U.S. Dept. of Commerce, Bureau of the Census.

Cleveland, W.S., Dunn, D.M. and Terpenning, I.J. (1978b), "The SABL seasonal adjustment package-statistical and graphical procedures", available from Computing Inf. Library, Bell Laboratories, Murray Hill, N.J.

Cox, D.R. and Lewis, P.A.W. (1966), The Statistical Analysis of Series of Events, Methuen, London.

Denby, L. and Martin, R.D. (1979), "Robust estimation of the first-order autoregressive parameter", Jour. Amer. Statist. Assoc., 74, 140-146.

Durbin, J. (1960), "The fitting of time series models", Rev. Int. Inst. Stat., 28.

Geweke, J. and Meese, R. (1979), "Estimating distributed lags of unknown order", unpublished report, University of Wisconsin-Madison.

Gnanadesikan, R. (1977), Methods for Statistical Data Analysis of Multivariate Observations, Wiley, New York.

Hampel, F.R. (1971), "A general qualitative definition of robustness", Annals Math. Stat., 42, 1887-1895.

Huber, P.J. (1964), "Robust estimation of a location parameter", Annals Math. Stat., 35, 73-101.

Huber, P.J. (1977), Robust Statistical Procedures, Soc. for Indust. and Applied Math., Philadelphia.

Jones, R.H. (1974), "Identification and autoregressive spectrum estimation", Tech. Rep. No. 6, Dept. of Comp. Sci., State Univ. of New York at Buffalo, Amherst, N.Y.

Kazakos, P., and Gray, R. (1979), "Robustness of estimators on stationary processes", Annals of Prob., 7, 989-1002.

Kleiner, B., Martin, R.D. and Thomson, D.J. (1979), "Robust estimation of power spectra", Jour. Roy. Stat. Soc., B , 41, 313-351.

Lamperti, J. (1977), Stochastic Processes: a Survey of the Mathematical Theory, Springer-Verlag, New York.

Luenberger, D. (1969), Optimization by Vector Space Methods, Wiley, New York.

Makhoul, J. (1975), "Linear prediction: a tutorial review", Proc. IEEE, 63, 561-580.

Martin, R.D. (1979a), "Robust estimation of time series autoregressions", in Robustness in Statistics, edited by R.L. Launer and G. Wilkinson, Academic Press, New York.

Martin, R.D. (1979b), "Approximate conditional-mean smoothers and interpolators", Proc. of Heidelberg Workshop on Curve Smoothing, T. Gasser and M. Rosenblatt, editors, Springer-Verlag, New York.

Martin, R.D. (1980a), "Robust estimation of autoregressive models", in Directions in Time Series, edited by Brillinger et al., Instit. of Math. Stat. Publication.

Martin, R.D. (1980b), "Robust methods for time series", Proc. 2nd Applied Time Series Symp., Tulsa, OK, edited by D. Findley, Academic Press (to appear).

Martin, R.D. (1980b), "Robust methods for time series", Nottingham Int. Time Series Conf., March, 1979 (to appear).

Nelson, C.R. (1973), Applied Time Series Analysis, Holden-Day, San Francisco.

Parzen, E. (1974), "Some recent advances in time series analysis", IEEE Trans. Auto. Control, 19, 723-730.

Phadke, M.S. and Kedem, G. (1978), "Computation of the exact likelihood function of multivariate moving average models", Biometrika, 65, 511-519.

Robinson, E.A. (1967), Statistical Communication and Detection, Griffin, London.

Shibata, R. (1976), "Selection of the order of an autoregressive model by Akaike's information criterion", Biometrika, 63, 117-126.

Thomson, D.J. (1977), "Spectrum estimation techniques for characterization and development of WT4 waveguide - I", Bell System Tech. Jour., 56, 1769-1815.

Tukey, J.W. (1960), "A survey of sampling from contaminated distributions", in Contributions to Probability and Statistics, edited by I. Olkin, Stanford Univ. Press, Stanford.

Tukey, J.W. (1977), Exploratory Data Analysis, Addison-Wesley, Reading, Mass.

Tukey, J.W. (1980), "Can we predict where 'time-series' should go next?", in Directions in Time Series, edited by Brillinger et al., Instit. of Math. Statist. publication.

Welch, P.D. (1967), "The use of FFT for estimation of power spectra: a method based on time averaging over short, modified periodograms", IEEE Trans. on Audio and Electroacoustics, 15, 70-73.

Welch, R.E. (1979), "Robust regression methods", Proc. of I.S.I. Meeting, Philippines.

Wold, H. (1954), A Study in the Analysis of Stationary Time Series, 2nd ed., Almqvist and Wiksell, Stockholm.

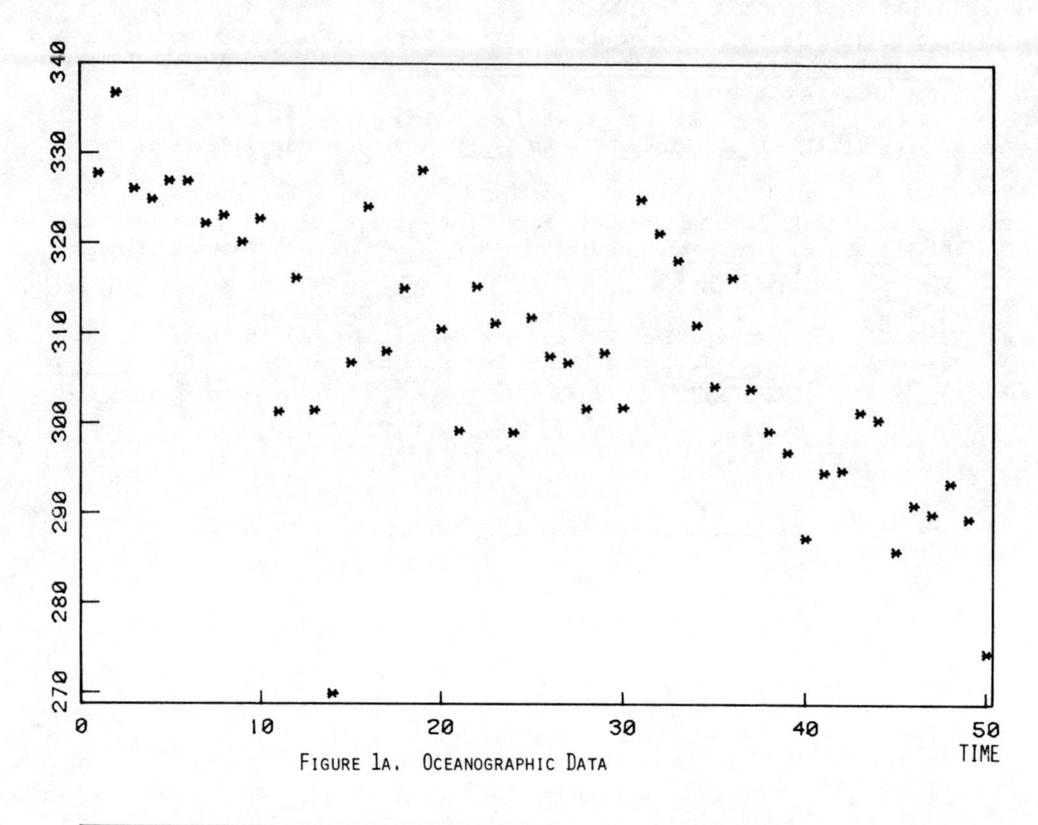

FIGURE 1A. OCEANOGRAPHIC DATA

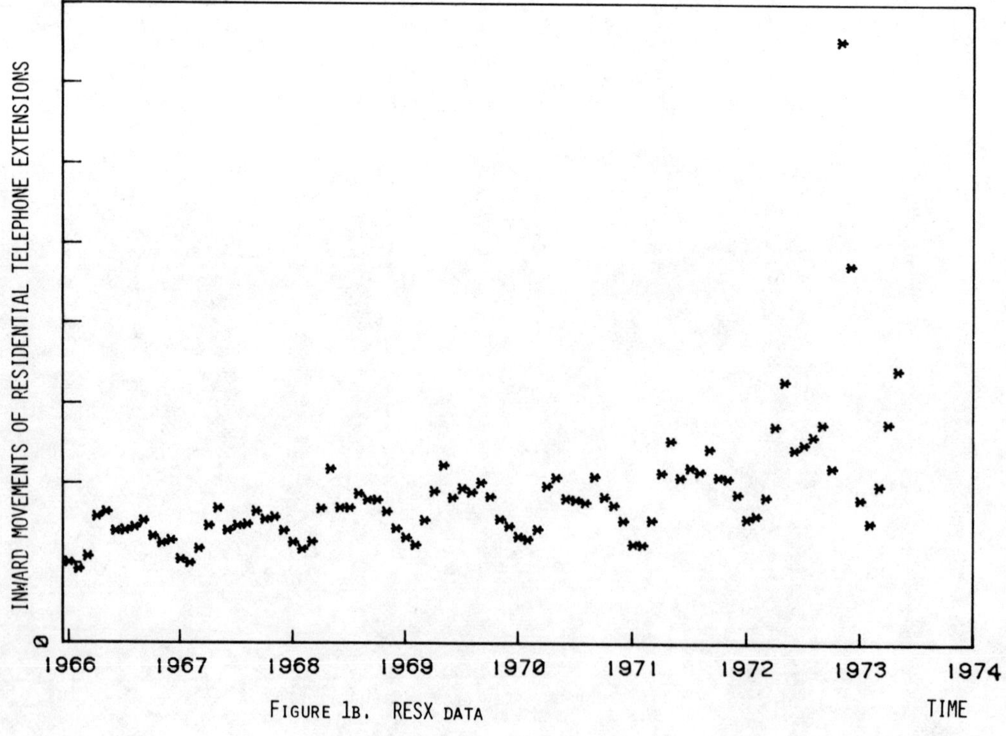

FIGURE 1B. RESX DATA

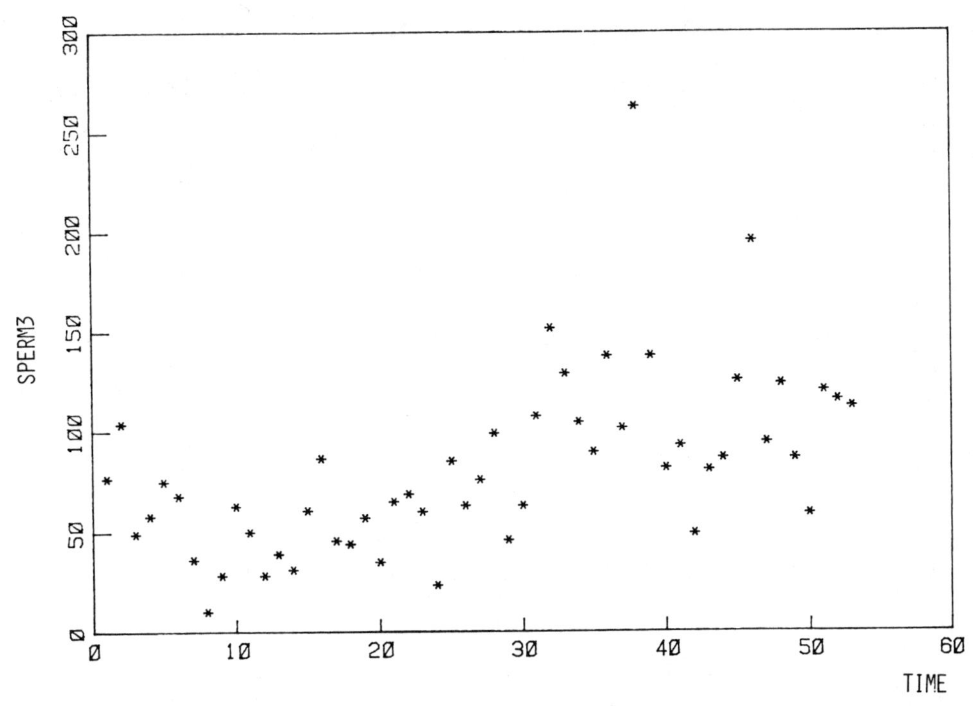

FIGURE 2. WEEKLY MALE SPERM COUNTS

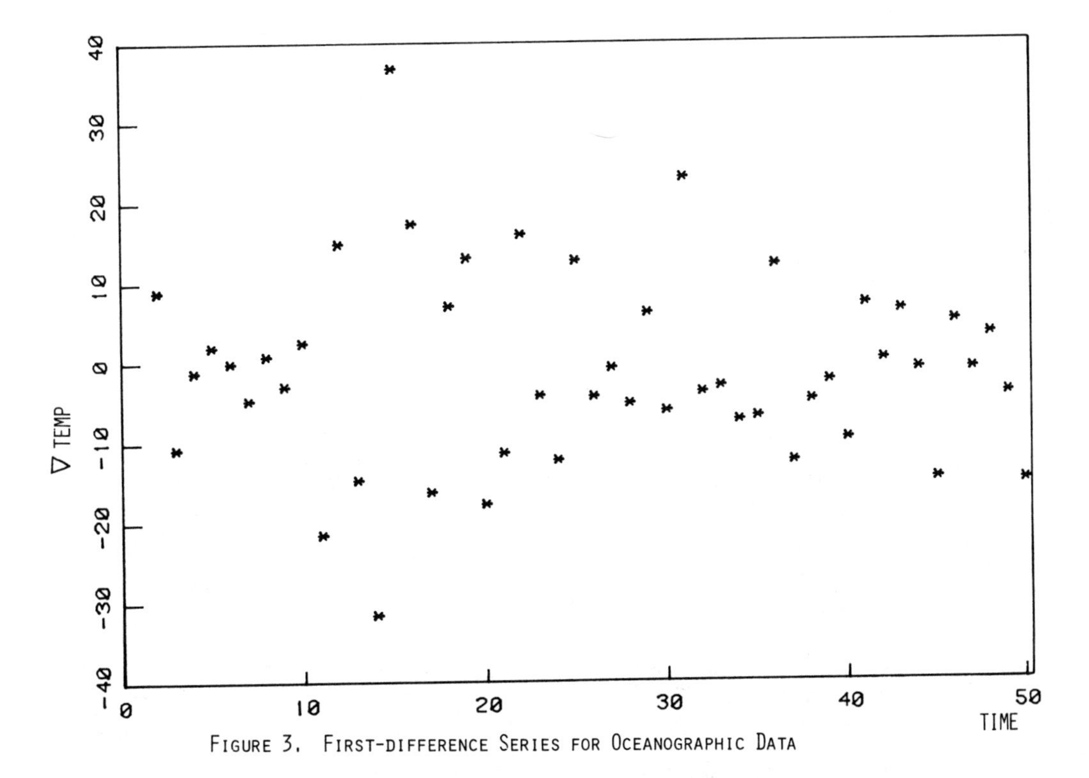

FIGURE 3. FIRST-DIFFERENCE SERIES FOR OCEANOGRAPHIC DATA

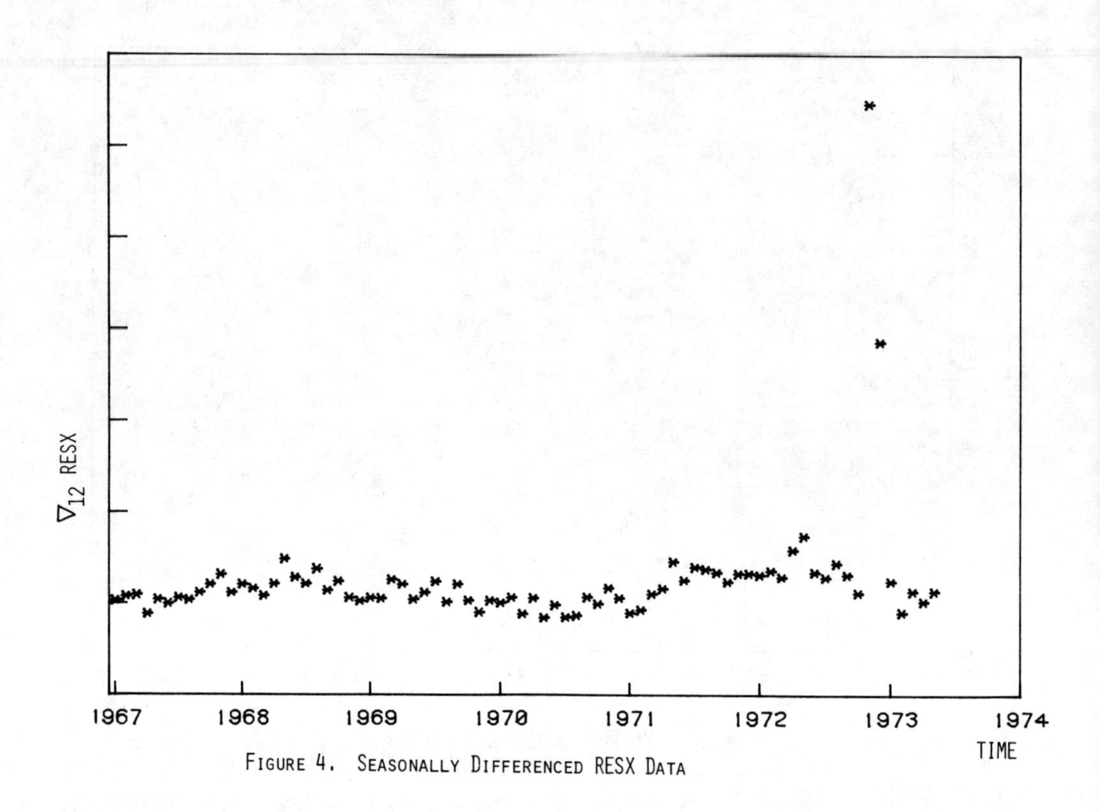

FIGURE 4. SEASONALLY DIFFERENCED RESX DATA

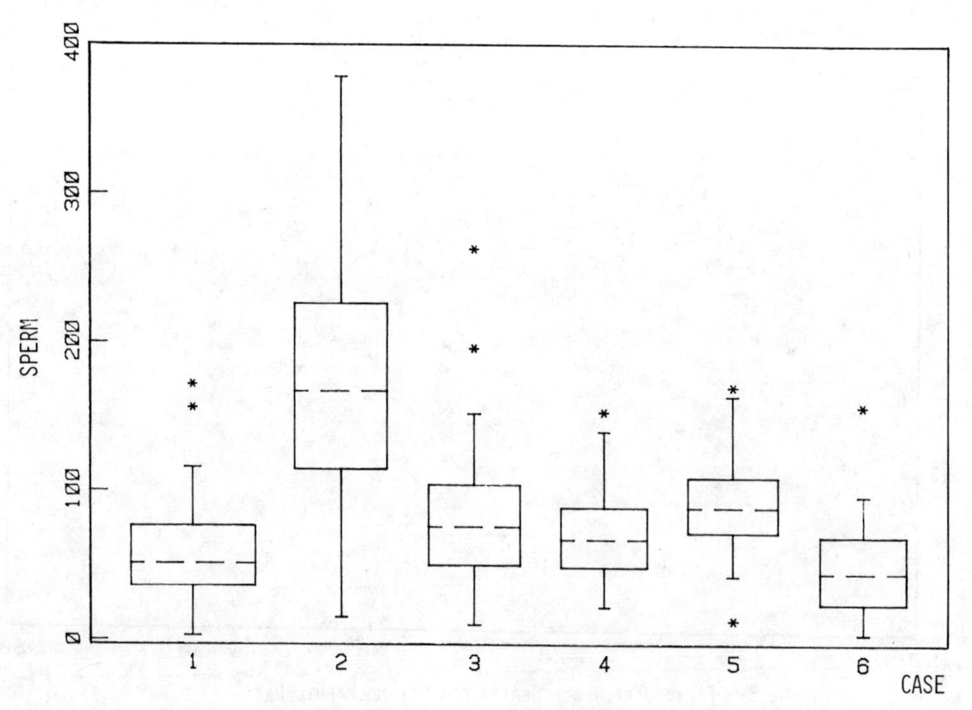

FIGURE 5. BOX-PLOTS OF SPERM COUNT DATA

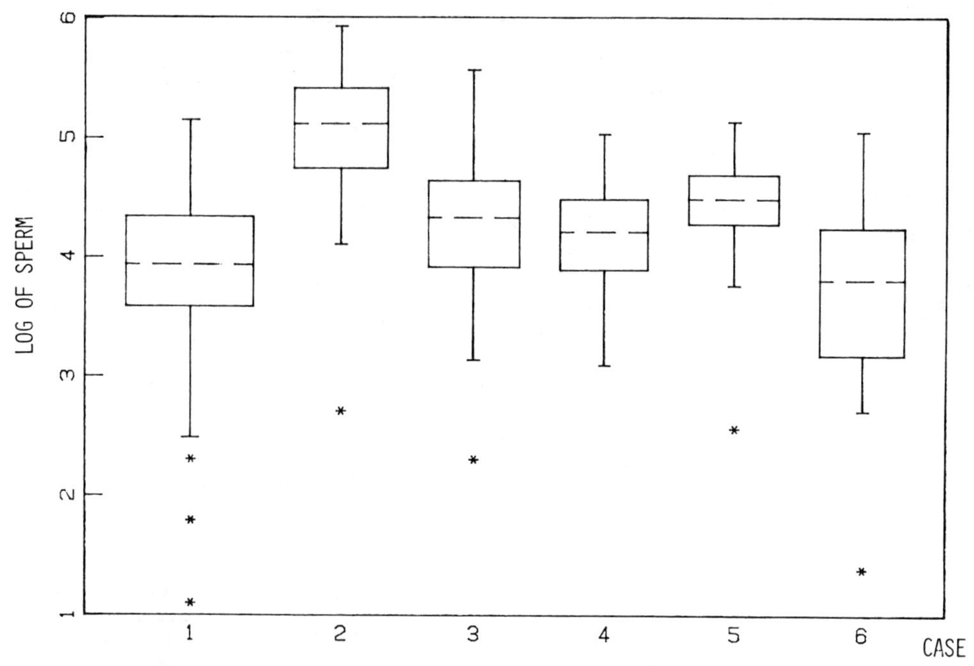

FIGURE 6. BOX-PLOTS OF LOGS OF SPERM COUNT DATA

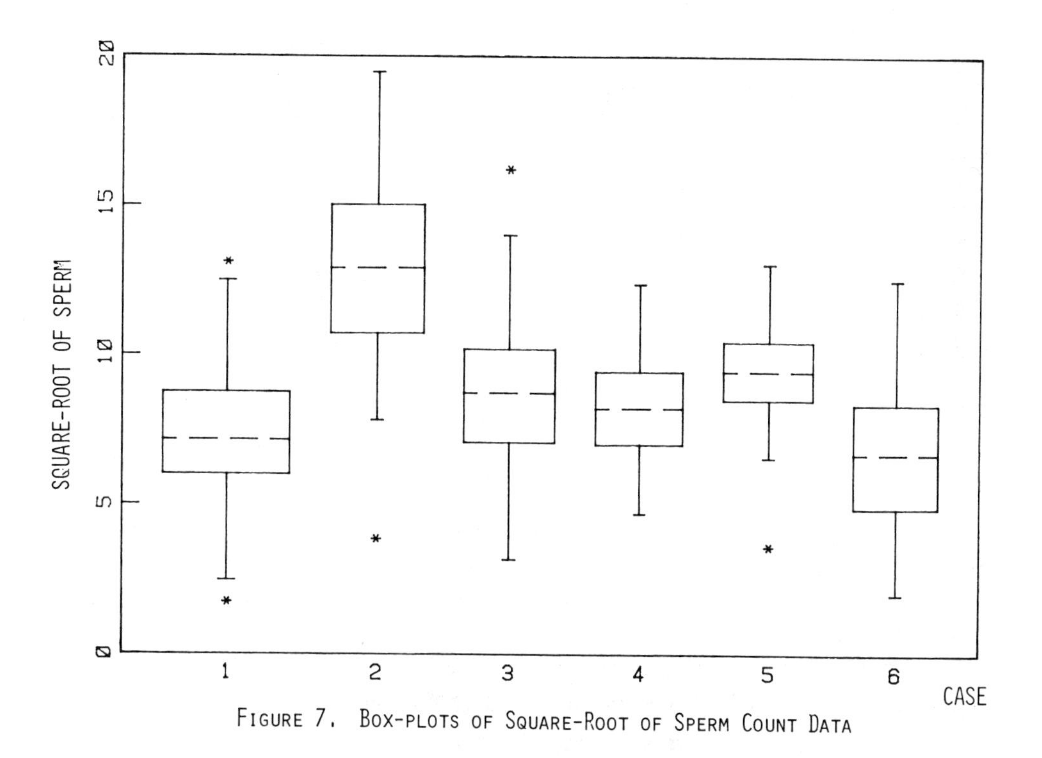

FIGURE 7. BOX-PLOTS OF SQUARE-ROOT OF SPERM COUNT DATA

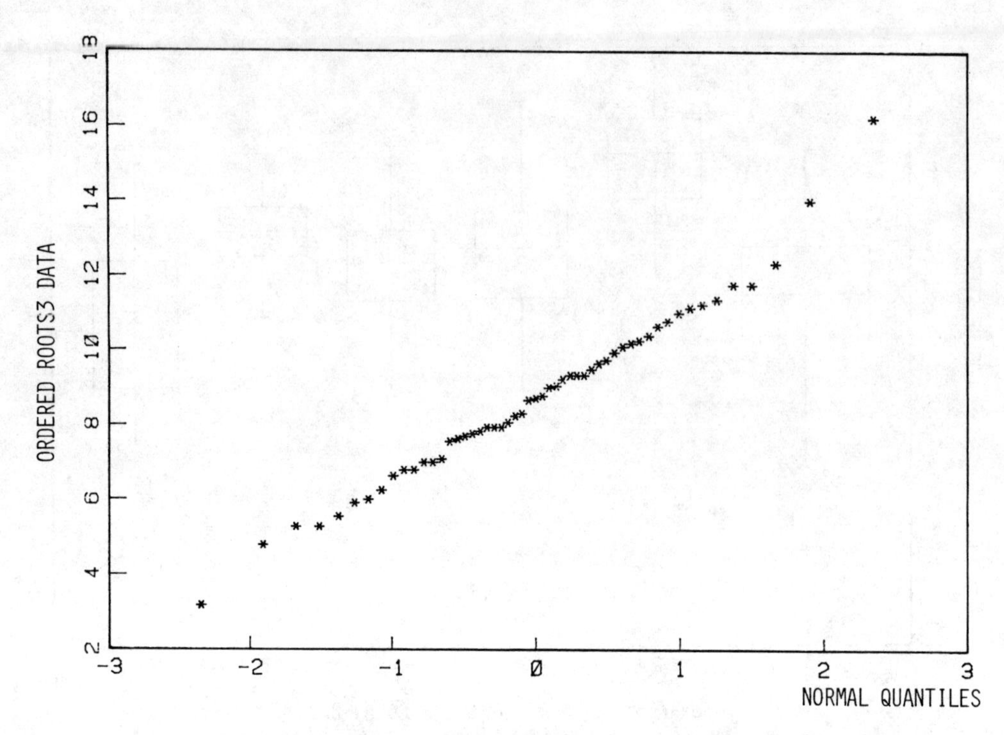

FIGURE 8. NORMAL PROBABILITY PLOT OF SPERM3.

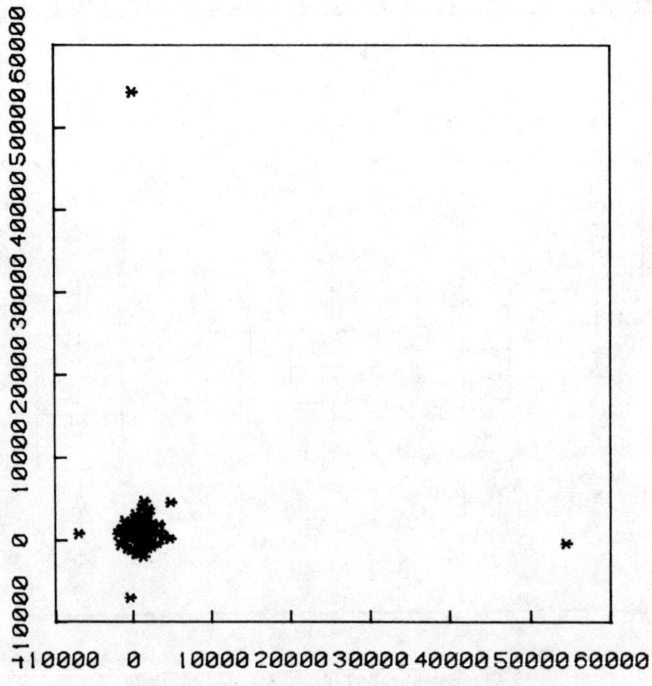

FIGURE 9. RESIDUALS VERSUS LAGGED RESIDUALS FOR LEAST-SQUARES AR(2) FIT TO ∇_{12} RESX DATA

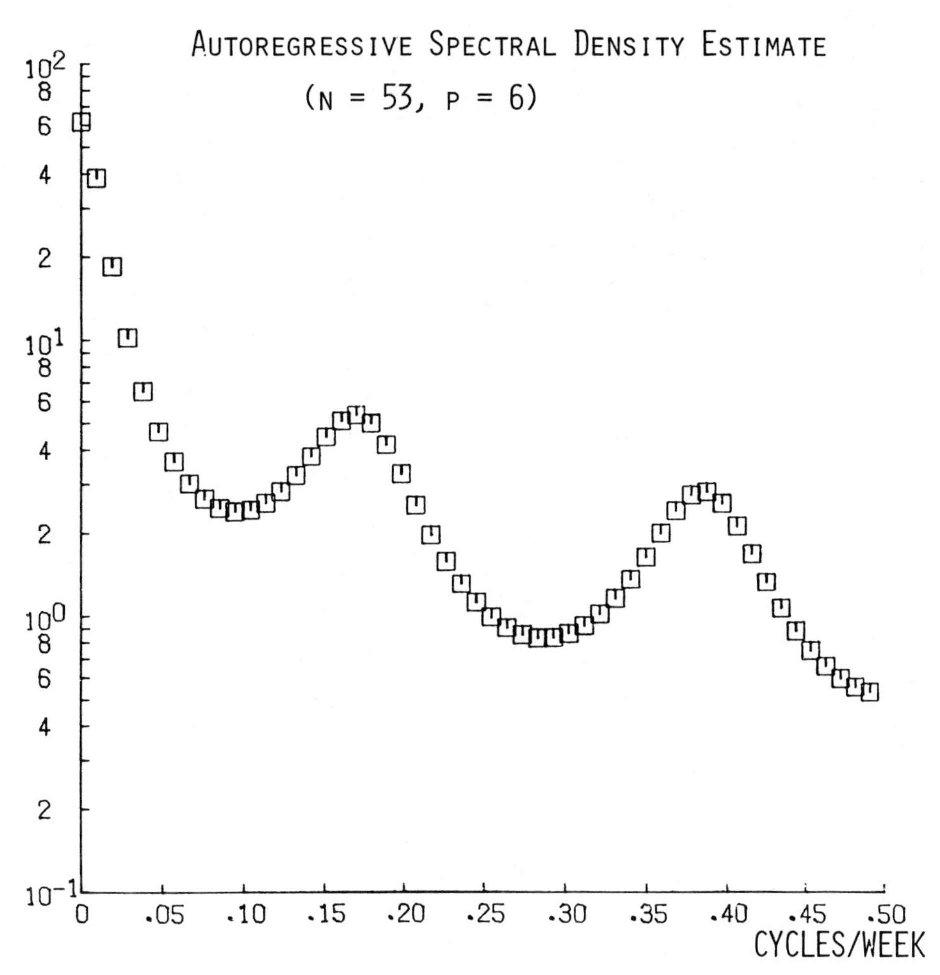

FIGURE 10. LEAST-SQUARES AR(6) SPECTRAL DENSITY ESTIMATE FOR SQUARE-ROOT OF SPERM3 DATA

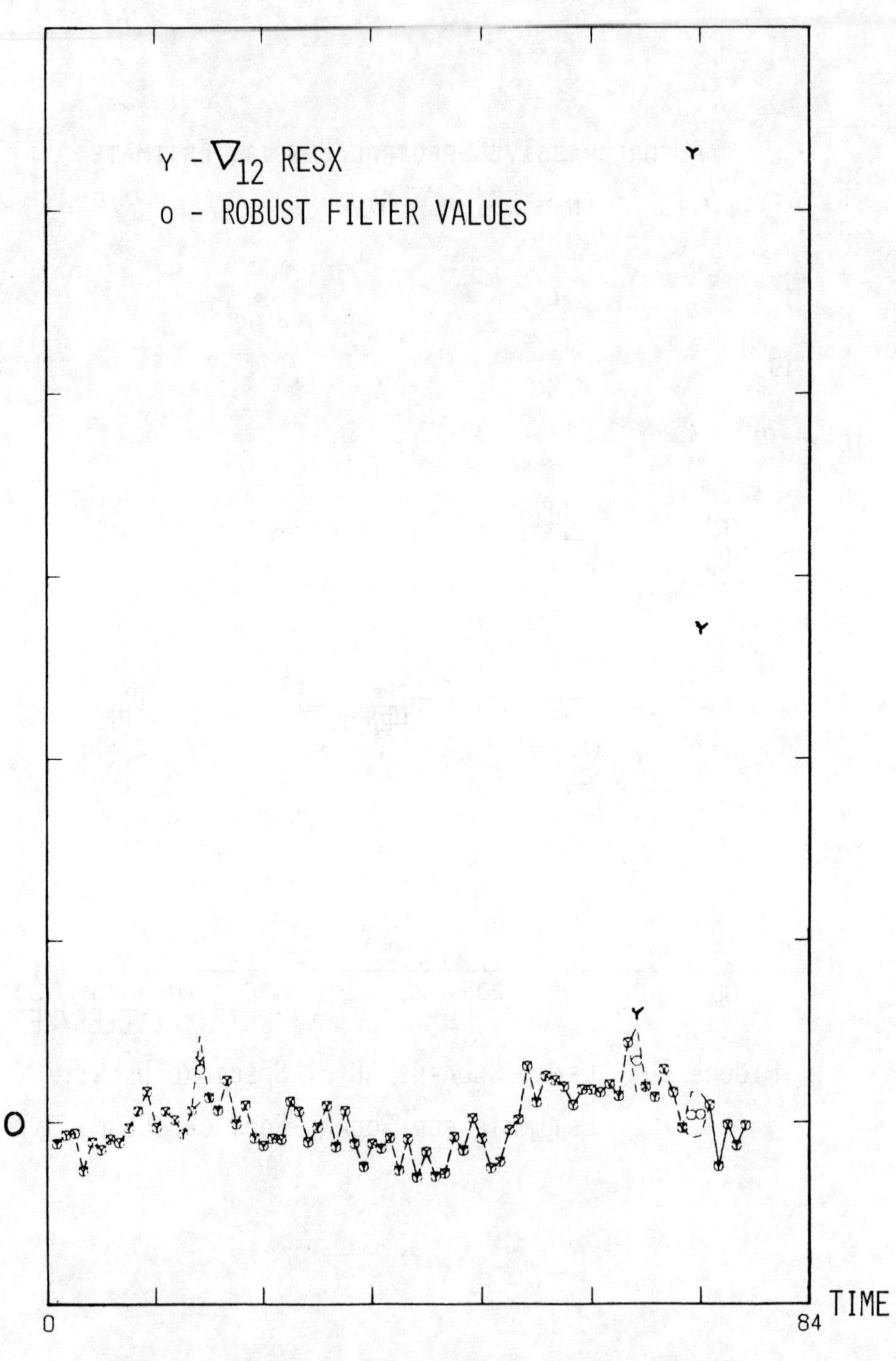

FIGURE 11. ORIGINAL AND ROBUSTLY FILTERED ∇_{12} RESX DATA

ABCDEFGHIJ–CM–89876543210